Trends and Developments in Modern Applications of Polyaniline

Edited by Florin Năstase

Published in London, United Kingdom

Trends and Developments in Modern Applications of Polyaniline
http://dx.doi.org/10.5772/intechopen.105245
Edited by Florin Năstase

Contributors
Jimmy J. Daka, George Mukupa, Hari Giri, Timothy J. Dowell, Mohammed Almtiri, Colleen N. Scott, Ozlem Erol, Lihi Abilevitch, Limor Mizrahi, Gali Cohen, Shmuel Kenig, Elizabeth Amir, Anita Grozdanov, Perica Paunović, Iva Dimitrievska, Aleksandar Petrovski, Gobind Mandal, Jayanta Bauri, Debashish Nayak, Sanjeev Kumar, Sarfaraz Ansari, Ram Bilash Choudhary

First published in London, United Kingdom, 2023 by IntechOpen
IntechOpen is the global imprint of INTECHOPEN LIMITED, registered in England and Wales, registration number: 11086078, 5 Princes Gate Court, London, SW7 2QJ, United Kingdom

British Library Cataloguing-in-Publication Data
A catalogue record for this book is available from the British Library

Additional hard and PDF copies can be obtained from orders@intechopen.com

Trends and Developments in Modern Applications of Polyaniline
Edited by Florin Năstase
p. cm.
Print ISBN 978-1-83769-616-1
Online ISBN 978-1-83769-615-4
eBook (PDF) ISBN 978-1-83769-617-8

Meet the editor

Florin Năstase is a senior research physicist at the National Institute for Research and Development in Microtechnologies, IMT Bucharest, Romania. In 2003, he obtained his master's degree in Polymer Physics at the University of Bucharest. He also obtained a Ph.D. (summa cum laude) in Physics from the same university in 2007 for his work on functional molecular architectures developed on π-conjugated oligomer/polymer structures. His current research interests include the synthesis of ultra-thin films using methods such as atomic layer deposition and plasma polymerization for a wide variety of applications, mostly within the areas of organic semiconductors, high-k dielectrics, ferroelectrics, nanocomposites, nanolaminates, and hybrid thin films. He is the founder of the Polymer Science Group, University of Bucharest, Romania, and a member of the organizing committee of ROCAM (Romanian Conference on Advanced Materials).

Contents

Preface

Synthetic polymers or plastic materials, whose history extends back 150 years, have numerous technical, medical, and scientific applications in industrial fields.

The high impact and success of these materials are typically associated with their wide range of characteristics/properties. For instance, they possess excellent combinations of physical and chemical properties such as high mechanical strength, toughness/impact resistance, flexibility, malleability, and low density, as well as good resistance to corrosive chemicals (solvents, acids, or bases), which are frequently incompatible with metals. They are also easy to synthesize and process, showing rapid applicability to the industry at low costs.

In general, plastics are associated with good electrical resistivity and are commonly employed as insulation for electrical cables. Recently (in the last 50 years), polymers have been associated with conducting or semiconducting applications, which confirms the proficiency to synthesize polymers with intrinsic conduction.

The first semiconducting polymer was reported in 1977. It was doped polyacetylene, which has electrical conductivity comparable to that of metals. The impact of this discovery did not go unnoticed. The three chemists who contributed to its development, Alan J. Heeger, Alan G. MacDiarmi, and Hideki Shirakawa, received the 2022 Nobel Prize for Chemistry.

Starting from this discovery, the interest in polymeric structures of "synthetic metals" has increased. In the last 50 years, the aim of many worldwide studies has been to easily develop environmentally friendly polymers with semiconducting or conducting properties specific to classic metals and semiconductors at low cost.

If the first-generation intrinsic conducting polymers show low processability due to low solubility and instability in the environment, the latest generation is easily processed in powders, flexible thin films, nanowires, and nanoparticles, being soluble in a variety of solvents and having high stability in the environment.

Although it was discovered more than 150 years ago, only recently has polyaniline received increased interest among the scientific community due to its intrinsic conduction properties, highlighting the possibility of polymer synthesis with high electrical conductivity.

Easy synthesis, moisture stability, and simple doping/de-doping chemistry are characteristics that distinguish polyaniline from all other conducting polymers. If systematic research has made possible the development of simple synthesis methods for polyaniline, additional efforts should be devoted to the complete understanding of the complex polymerization mechanism and its oxidation nature. With its rich chemistry and wide applicability, polyaniline is one of the extensively investigated polymers.

Usually, polyaniline is synthesized by either chemical or electrochemical oxidation of the aniline monomer in an acidic solution, and the aqueous medium is preferred. Although chemical and electrochemical synthesis methods of polyaniline are simple, requiring only some elements of fine chemistry, the polymerization mechanism is much more complex.

In addition to the large variety of chemical and electrochemical methods of synthesis, polyaniline has also been synthesized by unconventional methods, such as plasma polymerizations, vapor-phase deposition polymerizations (VDPs), inverse emulsion polymerization, photochemical polymerization, enzymatic polymerization, and autocatalysis polymerizations.

The synthesis of polyaniline by chemical, electrochemical, or other non-conventional methods is closely related to the desired final application. For instance, the electrochemical method is preferred when the final application requires highly ordered thin films. Plasma polymerization is the method of choice in the case of conformal, nanometer-thick films with high substrate adhesion that do not allow the use of solvents during the synthesis process.

Polyaniline is one of the most investigated and useful conjugated polymers with a dynamic evolution of applications. Assisted by both oxidation and protonation processes, it is the only conducting polymer whose electronic structure can be controlled in a reversible manner. In addition to the classical redox doping, polyaniline's unique electrochemical component also offers suitable chemistry for an easy doping/de-doping in the presence of "protonic acids" by which the number of electrons remains unchanged. It also exhibits intrinsic conductivity and a remarkable ability to undergo reversible changes from a conducting form to a semiconducting form and to a dielectric form. Last but not least, polyaniline is easy to synthesize and shows high stability in the environment.

Although many significant technological advances have been undertaken since the discovery of polyaniline's synthesis and processing, it is generally accepted that it cannot compete with traditional metals in applications such as electrical or transmission cables.

One of the first applications of polyaniline was in the structure of light batteries. Batteries have been developed with respectable performance using polyaniline as the positive electrode (cathode) and lithium-aluminum alloys as the negative electrode (anode). Polyaniline has also been used in biomedical applications; biosensors have been developed starting with the immobilization of the enzymes on a conducting polyaniline matrix. The color change property associated with polyaniline in various oxidation/reduction states has found potential uses in electrochromic devices, such as smart windows. Electrical charges and conformations of the multiple oxidation states also make polyaniline highly promising for applications such as supercapacitors or actuators. Polyaniline, with its distinct responses to doping processes (acid/base), makes it an ideal option for applications in chemical vapor or liquid sensor structures.

The different colors of different forms of polyaniline, along with its doping/de-doping response and charges and conformations of multiple oxidation/reduction states, makes polyaniline highly promising for antistatic applications, electromagnetic interference shielding, anti-corrosive coatings, toxic metal recovery, fuel cells, and microbial fuel cells.

Polyaniline has come a long way from a pure scientific laboratory curiosity to a material that can find its final destination in a wide range of commercial products.

My thanks go to all the authors who contributed to the writing of this book. I also wish to thank Ms. Zrinka Tomicic at IntechOpen for her support throughout the publication process.

I dedicate this book to my daughter, Ana Maria, who gave me time.

Florin Năstase
National Institute for Research and Development in Microtechnologies - IMT
Bucharest,
Bucharest, Romania

Chapter 1

Polyaniline Nanostructures: Techniques in Structure-Tailored Polymerisation-Superstructures

Jimmy J. Daka and George Mukupa

Abstract

Polyaniline (PANI) is one of the widely studied conducting polymers. As such it is one of the widely applied conducting polymers for laboratory bench work applications. The limitation to application in commercial work has been hampered by the inherent difficulty of a polymer being processed once synthesised. The solution to this lies in synthesising the PANI that has uniform structures ready for application in that form or creating a composite with other molecules that bring about the level of processability to acceptable processible levels. This paper seeks to outline the general synthetic underlying principles behind the synthesis of PANI that may bear the structural nature for ready-to-apply or processible to some extent for possible application. The paper outlines the general synthetic concept framework for one to manipulate for suited use.

Keywords: structure, tailored, synthesis, techniques, polyaniline

1. Introduction

1.1 Polyaniline as conductor polymer, historical perspective and synthetic techniques

In this chapter, structured tailored polyaniline (PANI) synthesis schemes will be focused on, the body will provide various synthetic schemes that may result in a specifically shaped PANI. The structured tailored PANI is seen as ideal for applications.

Polyaniline (PANI) is one of the most highly studied intrinsic conductor polymers (C.P). This has been due in part to its conductivity [1], stability of polymer [1, 2] and ease of polymerisation process [1–6]. Intrinsic conductor polymers are organic polymers with the ability to allow electricity to flow through them or they allow the transfer of charge from one point to another in the polymer chain. In particular, they possess the conjugations in the chain that may alter the energy gap and electronegativity through the various forms of synthesis. For polyaniline-based conductor polymers, in principle, the chain is made up of benzenoid and quinoid linked through the amine [6–13]. The amine could be protonated and deprotonated by the use of acid

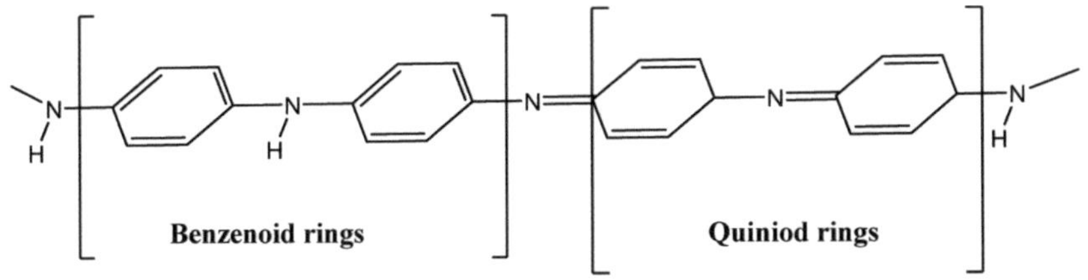

Figure 1.
The benzenoid and quinoid are the basic alignments of aniline monomers in polyamine [13, 17].

enrich or depleting substances called mainly referred to as dopants [1, 12–17]. The diagram in **Figure 1** shows the basic backbone polyaniline is the simplest form.

The number of benzenoids and quinoids in the chain gives the distinctive colour of different PANIs that have been reported. The basic colours reported so far include emeraldine green, violet, purple and colourless. The colouration can be structurally understood, as given in **Figure 2**.

1.2 A brief history of polyaniline (PANI)

The earliest moments of researchers working with polyaniline or its predecessor named black dye, dates back to the 1800s [18, 19]. The present name of polyaniline was identified in the 1960s, earlier to which it was named based on various colour-based names and commonly referred to as aniline black. The colour-based name code was mainly because the primary application of early polyaniline was as green, blue and black dyes for early cotton, with industrialisation in full swing, in the early days of industrialisation, dyes date far as the 1800s [19, 20]. The current momentum and interest in PANI have grown with the after-match of the 2000 Nobel Prize and subsequent accelerated desire for conductor polymers and their application [21, 22]. Generally, the drawback to PANI application is its inability to be processed after it has been polymerised, the rigidity of a polymer and unpredictable nanostructures from the easy methods to polymerise it that is chemical oxidative and electrochemical techniques.

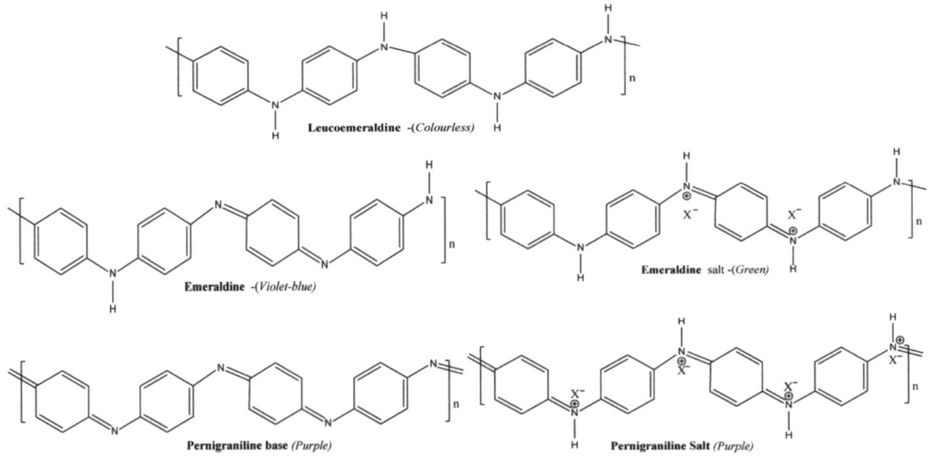

Figure 2.
Shows different forms of polyaniline with the expected colour.

1.3 Polyaniline (PANI) synthesis techniques

Polyaniline is generally synthesised through oxidative polymerisation techniques. The process involves polymerisation where monomers usually aniline or derivatives of aniline in the presence of the oxidising agent or electric field in an acidic medium [19, 20, 23–27]. Specific techniques such as (i) chemical oxidative polymerisation (ii) electrochemical polymerisation, (iii) plasma polymerisation (iv) enzyme catalysed polymerisation, (v) bio mimic catalysed polymerisation and (v) microwave-assisted polymerisation [6, 8, 10, 28–32]. The undertaking is done to control the type of shapes.

2. Polyaniline: structure-tailored synthesis for possible application

Structure-tailored polymerisations are the methods used to control the molecular structures to obtain the structurally predictable, consistent properties of PANI. This is generally done by adjusting reaction or polymerisation conditions to direct the reaction towards the desired goal. Various studies have been conducted when it comes to control of morphology, crystallinity and conductivity of polyaniline [10, 29, 33, 34]. What this means is that if PANI with a desired and targeted application is to be made, then it means implementing a scheme of synthesis designed with the specific outcome of structure-tailored PANI in the back of the synthesis plan. The following techniques are used to obtain structure-tailored PANI.

- Template-assisted synthesis (hard and soft templates) [33, 35].
- Molecular imprinting [10, 36–41].
- Dopant-assisted polymerisation [42–48].
- Emulsion-based (inverted and non-inverted) techniques [48–55].
- Composite polymerisation [56–61].

2.1 Polyaniline synthesis template-assisted synthesis: soft templates

Soft templates are mainly solutions with the ability to line up the acidified or non-acidified aniline monomers. The soft templates tend to have multiple functions in PANI production. Among the roles, it may work as a surfactant for dispersing the solution of anilinium, providing a local environment that tends to affect the pH of the immediate surrounding of the solution, composites matrix for much processible PANI and in some cases work as a doping material itself [62–66]. Chen was able to use the suprastructures of methyl orange in a low pH solution. It was established that the effectively self-assembled supramolecular aggregation in the shape of flake and dendrite could work as soft templates for the preparation of micro-tubes of PANI the monomer was aniline the activating agent for the reaction was ammonium peroxydisulphate APS [35]. Meanwhile, Qui et al. were able to synthesise highly crystalline PANI nanofibres and nanorods by changing the concentration of sucrose octaacetate, 2.0 g and 3.0 g, respectively [62]. The solution of sucrose octaacetate was working as a soft template. It was observed that they were significantly irregularly

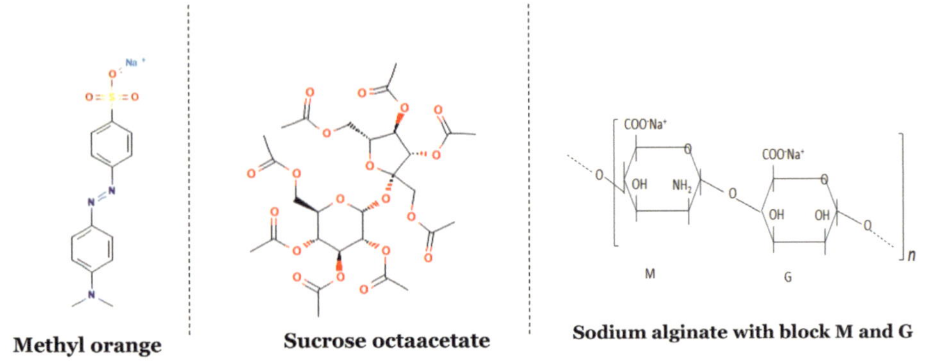

Figure 3.
Showing chemical structures of molecules used in the soft template: Methyl orange, sucrose octaacetate and sodium alginate [64–66].

shaped agglomerates and scaffolding formed at high concentrations. Chattopadhyay et al. [63], synthesised structure-tailored PANI by oxidative polymerisation of aniline monomers in the presence of sodium alginate. They were able to produce nanofibres, ammonium peroxydisulphate was used as an oxidising agent while sodium alginate worked as a soft template. The reaction was carried out in the presence of hydrochloric acid. For soft templates, the concept is that a molecule with some solubilities in an aqueous medium helps align the aniline or anilinium ions in the solution. When close examination of the molecules so far used, will show they show some centre in the molecule with the potential to be ionised in an acidic or basic medium. **Figure 3** shows molecules of methyl orange, sucrose octaacetate and sodium alginate.

The monomer is still aniline, which in an acidic medium, is protonated and becomes anilinium ion. The molecules used in aqueous medium help align the anilinium ion just before oxidising agent can activate the process of oxidative polymerisation [8, 67, 68]. The process can be understood in **Figure 4**.

Therefore, for the soft template, it could be understood as in **Figure 5**. The soft templates help in aligning the protonated aniline molecules which then get oxidised by APS or hydrogen peroxide or another oxidising agent that is usable for the reaction.

This is not much different from other types of soft templates. Their role is typically that of aligning monomers. But at other times, they too help in creating the background for the composite formation and subsequent doping or points where the pH of the local environment is more strongly acidic than the solution of the reaction medium [34, 68–70]. The soft templates in synthesis of structure-tailored polyaniline means that, the use of small molecules or polymers for controlling the general growth of a highly crystalline PANI. The soft templates may also serve other purposes such as doping the PANI, solubility hence processibility and easy of post-synthesis processing.

2.2 Polyaniline synthesis template-assisted synthesis: hard templates

Hard templates could generally be understood as any substance that is solid or gel-like material that helps in aligning the other materials. In this context, even composite forming materials that intertwine with the polyaniline forming on conducting matrix still could qualify to be referred to as hard templates [70, 71]. Wei et al., synthesised the enzyme-catalysed polyanionic template based on sulfonated polystyrene. Meanwhile, a hard template may entail fabricating polyaniline in the structures of hard

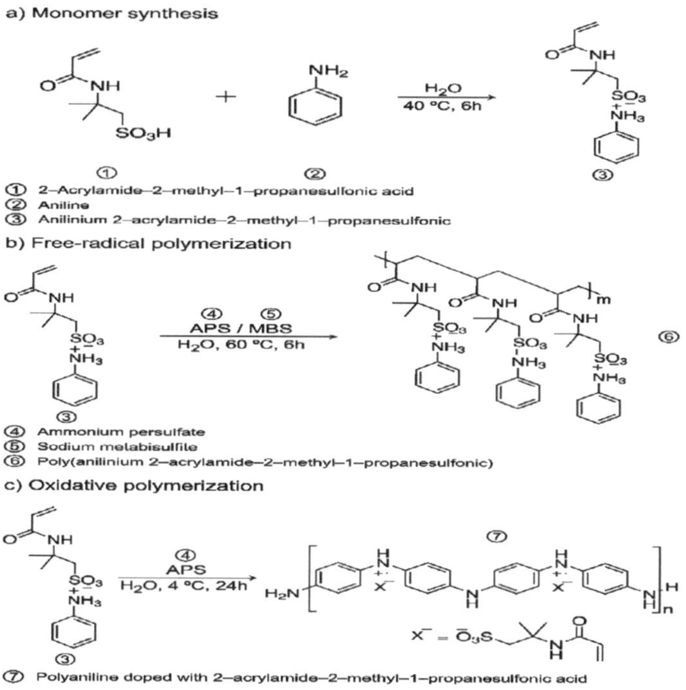

Figure 4.
Showing the anilinium taking part in the various reaction conditions and mechanisms [68].

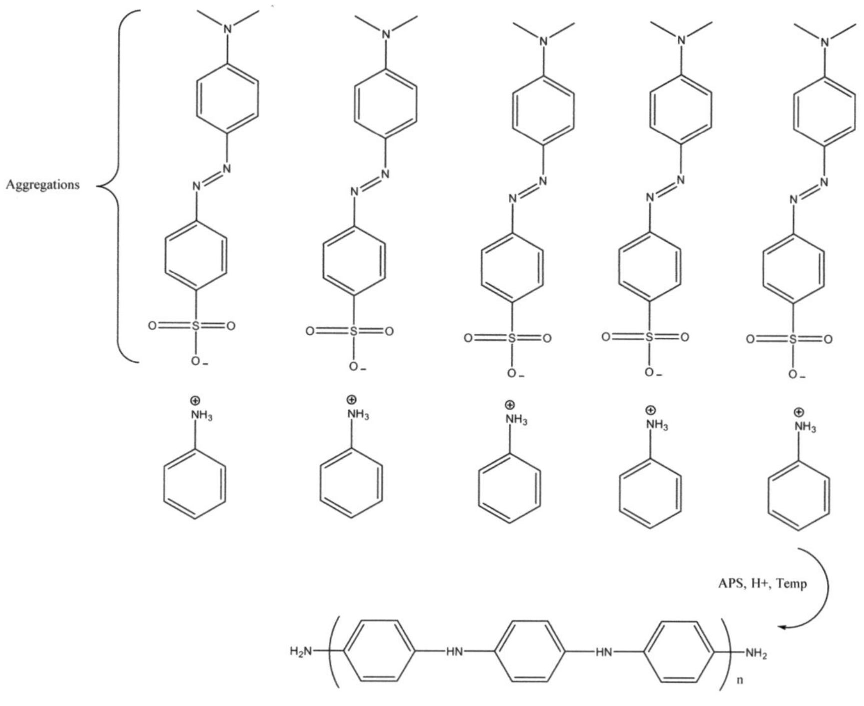

Figure 5.
Shows the methyl orange molecule being used as a soft template for the synthesis of PANI.

templates, this technique not only has high specificity, predictability and enhanced electro-activities of the PANI so synthesised, but it is easy and the throughput is relatively high [72, 73]. Other techniques in the structure-tailored polymerisation on hard templates include electrochemical polymerisation on the surface of charged hard plates, predetermined nanostructured material to resemble the PANI nanostructures, growth on the surface of other materials that have a specific design [74, 75].

Figure 6 shows electrochemical polymerisation using a conducting hard template mechanism. The electroactive surface can be graphene, metal plate or zeolite on a semi-conducting surface.

Then, for halloysite PANI synthesis, Lui et al. [33] used typical hard structure halloysite-suited materials that were used to make nanotubes, **Figure 7** depicts on concept.

The shapes of nanostructures obtained were hollow tube-like structures resembling the halloysite they were fabricated from. Their SEM images were remarkably shaped as the hard templates they fabricated from. **Figure 8** shows the actual produced PANI.

The halloysite templates provide surfaces that provided foundation and support for the nanostructures of PANI synthesised by the technique. The polymerisation technique remains anilinium and APS or any other oxidising material that may be suitable. **Figure 9** outlines the general concept of the halloysite technique for the synthesis of PANI.

2.3 Polyaniline synthesis: molecular imprinting synthesis

Molecular imprinting (MIP) is the technique used to create artificial receptor recognition sites which have an enhanced selectivity and specificity to a specific molecule usually referred to as a target molecule [10, 36, 37]. In MIPs, the synthesis involves the synthesising of a backbone-bearing polymer in this case polyaniline. The synthesis is simultaneously done in the presence of the template or substrate molecule, which is achieved with another molecule called a cross-linker [10, 38]. Then, the template molecules or substrates are removed from the backbone and cross-linked

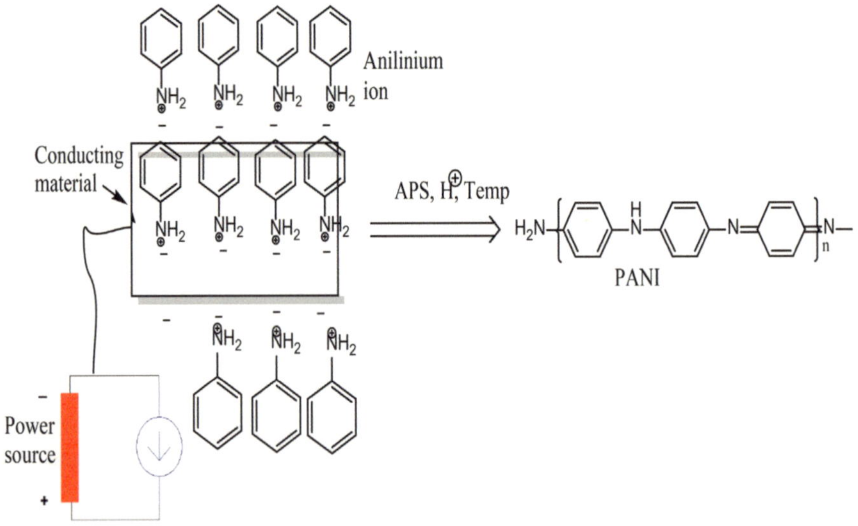

Figure 6.
Shows the technique of polymerisation using the conducting material as aligning surface.

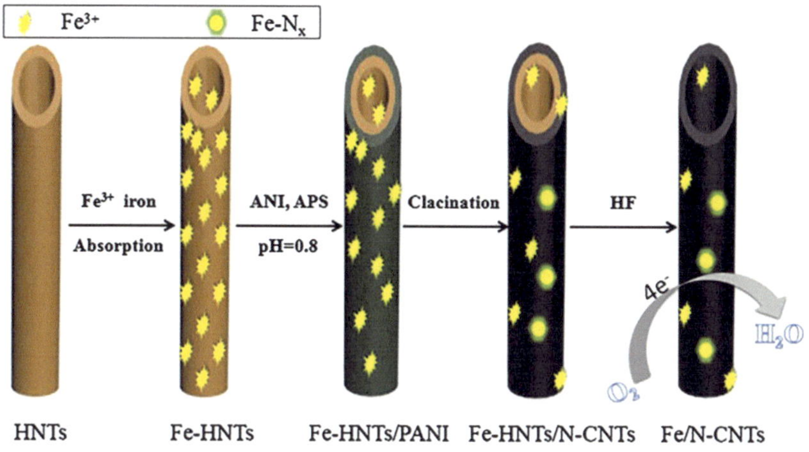

Figure 7.
Illustration of halloysite for PANI synthesis [33].

Figure 8.
Typical halloysite structure tailored PANI [33].

matrix. The process leads to cross-linked matrix polymers, which have cavities that are size, shape and functional group specific. It is this high specificity to a template or substrate molecule that brings about the tailor-structured polyaniline nanostructures. The polymer produced in the final state is not just specific at the molecular level in terms of complementary sites but is still the consistency of suprastructures.

Bagdžiūnas, synthesised what he termed a supramolecular system with tailor-made sites for binding, which stood complementary to molecule templates in each of the following parameter's size, shape and specific functional groups intending to evaluate its analytical, physical and theoretical interactions [38]. Different teams have reported nanospheres, nanofibres and nanotubes [39–41]. The synthesis technique may use any type of polymerisation available on PANI [38]. Therefore, the general principle could be understood in **Figure 10**.

The molecular imprinting (MIPs) technique can be performed following any other standard polymerisation types [10, 38–41]. In many instances for PANI due to

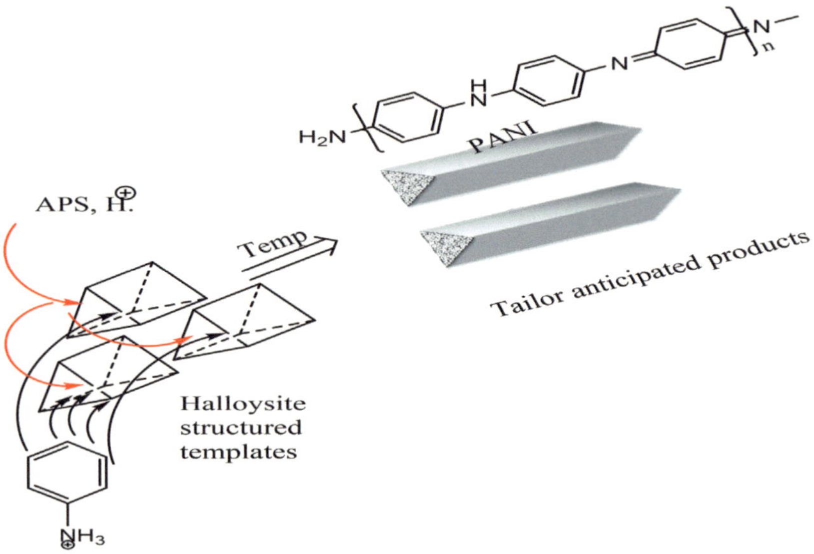

Figure 9.
The general concept of halloysite hard template for PANI synthesis.

its inherent difficulty of processing once polymerised, the in printing in molecular in printing techniques it is advisable to put in one mix the aniline monomers, oxidising agent, cross linkers and the substrate/template molecules. Based on the physical properties of a substrate/template molecules can be removed from the cavities [10].

2.4 Polyaniline synthesis: dopant-assisted synthesis

In a quest to synthesise structured tailored dopant-assisted synthesis has been used to try to control or manage the morphological makeup of the structures. In principle, dopant-assisted polymerisation is the method used in polymer synthesis by adding a small number of materials called dopants into the polymer to change monomer reactivity and influence the chain growth and structures of the polymer. The dopant used is referred to functional dopant. The functional dopant is meanwhile defined as molecules that are added to conductor polymers to enhance conductor polymers' conductivity [43]. Hafizah et al. [43], synthesised PANI nanoparticles whose particle sizes were manipulated by the amount of sodium dodecyl sulphate SDS [44]. The same SDS or SLS has been reported as playing the duo role in PANI synthesis that of the surfactant for dispersing the aniline or anilinium ions and the dopant material. Alshareef et al. were able to synthesise morphology-controlled PANI nanostructures based on altering three parameters; (i) tunable oxidant manganese (IV) oxide (MnO_2) as a reactive template, (ii) redox-active electrolyte and (iii) porousness of the synthesised PANI [46]. The other team made multi-layered morphology-controlled PANI. The key to control of morphology is the SDS that helped control the shape and ionic atmosphere where the PANI had a morphology [45]. It has also been established that SDS can work as a dopant and also a wettability control medium (processability) [47]. Hence, to prepare some morphologically controlled PANI, dopants or redox-active chemical reagents must be incorporated into the polymer final structure. The dopant selection for versatility's sake may need to have multiple roles in the synthesis scheme. They can follow any of the known synthesis protocols.

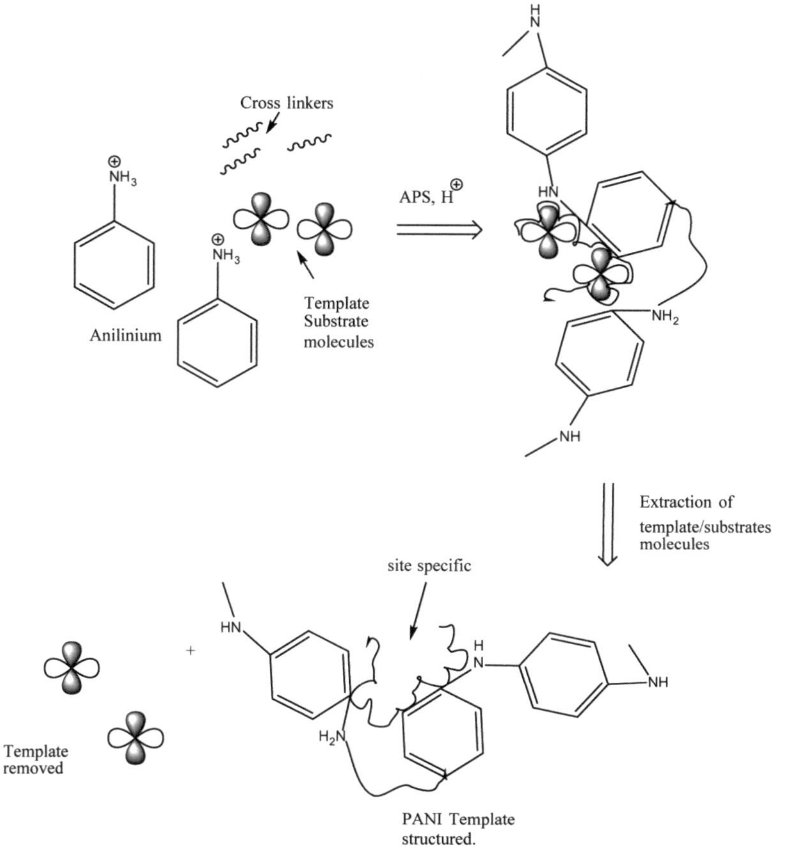

Figure 10.
Shows a simplified synthesis in molecular in printing structure tailored PANI.

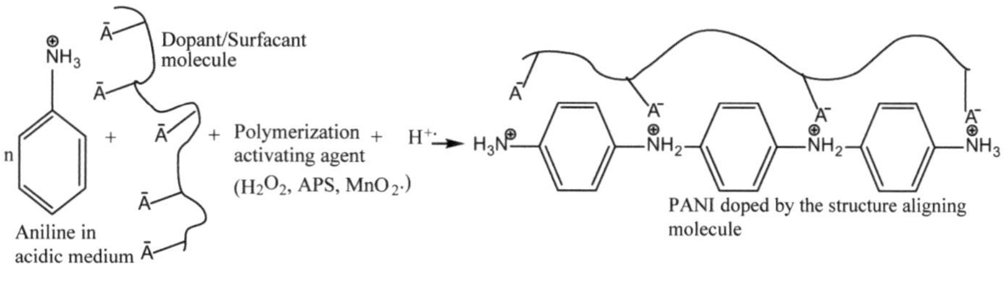

Figure 11.
The general understanding of dopant-assisted structure tailored PANI.

The general reaction equation could be understood in **Figure 11**.

Meanwhile, the specific cases where dopants were used as aligning molecules and dopants is Shen et al., and Goswami et al. in each where various substituted sulfonic acids were used in aligning oligomers for PANI synthesis for a particular application [76, 77]. The technique produced fibres of the polymers of PANI that were meant for the application. **Figure 12** shows the typical application involving the structures of PANI obtained.

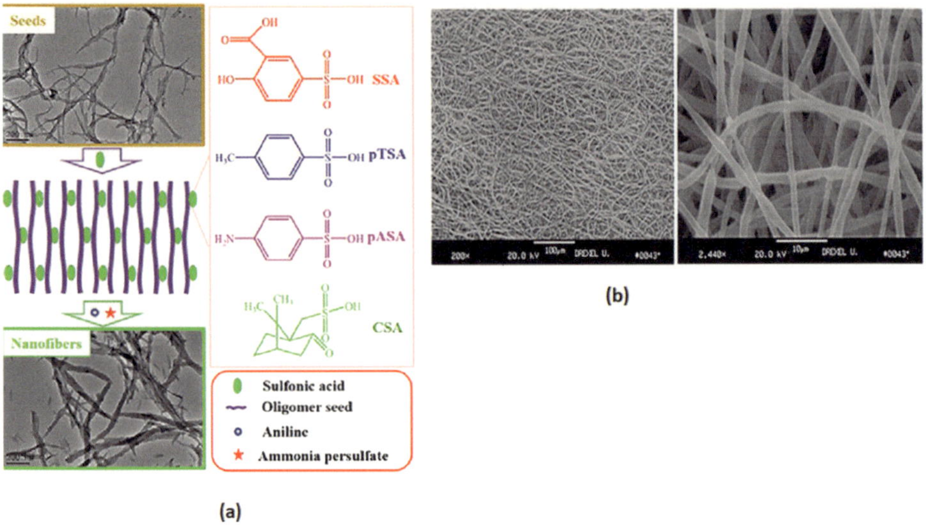

Figure 12.
(a) Dopants used in PANI for superconductor capacitors and (b) the SEM images for the nanofibres produced [76, 77].

2.5 Polyaniline synthesis emulsion as structure-tailored polymerisation technique

Emulsion polymerisation is the type of polymerisation technique that allows the formation of a polymer on the surfaces of the layers of two solutions that may have different densities (miscibility). The polymerisation is sometimes done in the presence of surfactants that provide stability to the reaction adduct. Emulsion polymerisation is widely applied in PANI synthesis with a view to controlling the narrowness of nanostructures, distribution of particles and reduce the residue monomers in the final product—oligomers and dimers [48]. In emulsion polymerisation, the monomer molecules are emulsified in a continuous phase of an immiscible liquid, while the activator is dispersed in another layer of an aqueous system. Then, the two solutions are brought in contact with each other and allowed to interact. Since the solutions are not miscible, they will layer up but at the boundary, the polymer will form [49]. The polymer and its oligomers may diffuse in a layer favourable to their polarity. Meanwhile, the basic synthesis of PANI is still maintained where you use aniline monomer, oxidising materials [8, 10]. The emulsion can be carried out in the following four major types of emulsion (i) microemulsion, (ii) macroemulsion, (iii) miniemulsion and (iv) inverse emulsion (inverted).

(i) Microemulsion is the type of emulsion where the dispersion is oil water and the particles of dispersion range from 1 nm to 100 nm, though common practice shows a range between 10 nm and 50 nm being practised in synthesis.

(ii) Macroemulsion is the type of emulsion where the dispersion is oil water and the particles are in the range of 1–100 μm.

(iii) Miniemulsion is the type of emulsion where the oil and water are the dispersing media, but the particle sizes range between 50 nm and 1 μm.

(iv) In inverted emulsion on the other hand the dispersion is water oil medium, especially with denser than oil/organic medium that is denser than water [50].

In any case, it is understood that the monomer is dispersed in one medium, while the activator or oxidants or initiator is set in another medium. Then, as the two molecules come in contact either of boundary or during transfer the polymer is building up at the interface of the two. Emulsion polymerisation by desire is meant to control the sizes of particles, distribution and uniformity of particles in the final product [48]. So, for PANI structured tailored suprastructures, the technique entails dispersing the aniline or anilinium ion monomers in an organic medium. Sometimes it may require stabilisation employing the addition of surfactants. Chajanovsky and Suckeveriene managed to synthesise PANI hybrid nanoparticles by inverse emulsion but used ethanol as the medium where aniline was dispersed, while APS was dispersed in water [51]. Österholm et al. reported synthesis of PANI through a weakly acidic medium but stabilised by dodecyl benzene sulphonic acid DBSA [52]. Perrin et al. reported a method of synthesis of PANI, through emulsion but the oxidant APS was dispersed in decylphosponic acid (DPA), while the aniline monomer was dispersed in a mixture of heptane and chloroform, it was found that the emulsion synthesised PANI ratio improved rate of polymerisation increased with the decrease in DPA concentration [53]. In principle, emulsion polymerisation may require aniline monomers, stabilising agents, oxidising agents and its polar solvent or medium for dispersing it. The general understanding of the PANI synthesis could be as shown in **Figure 13**.

In general emulsion polymerisation, oil is used as dispersion molecules for monomers [50]. However, for PANI, it is other polymeric solutions such as surfactants that work well. These include dodecyl benzene sulphonic acid, sodium dodecyl sulphate (SDS), sodium dodecyl benzene sulphonate and Triton X 100, [8, 10, 51–56, 78]. The oxidants are APS, hydrogen peroxide, peroxide salts, dichromate solutions, iron (III) oxide and manganese (IV) oxide in addition to any redox controlling agent.

2.6 Polyaniline synthesis: composite structures

Composite materials are materials that are made from a combination of two or more materials of different chemical and physical properties for synergistic abilities when combined [56]. For polyaniline suprastructures, it can be said these are structures that

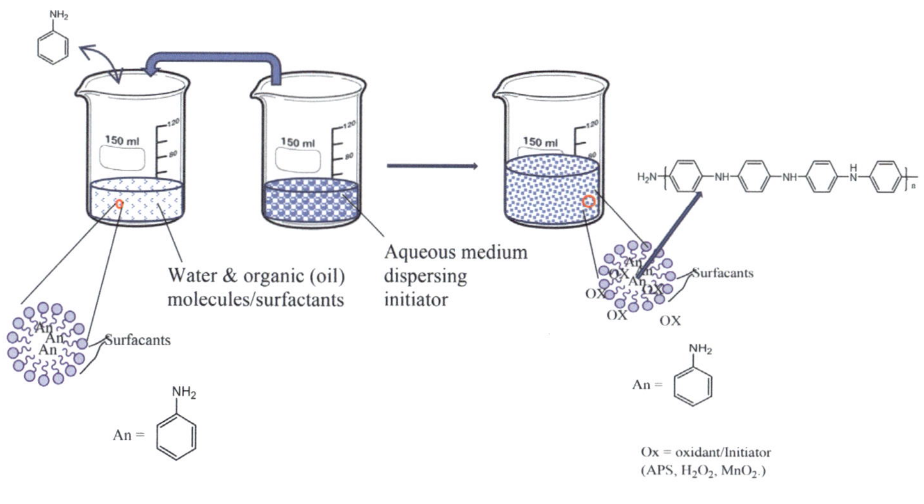

Figure 13.
Showing the general understanding of emulsion polymerisation for PANI.

are hierarchically organised mixtures of PANI with other materials such as nanotubes carbon, graphene, metals coreshell, polymeric molecules, clay, silicate, cellulose whiskers and different metal oxides [57]. The PANI composite nanostructures do not only have high nanoscale dispersing abilities but they possess far better properties than individual components of the composite. In order to obtain the structured tailored nanocomposite materials for PANI, they are three basic routes the synthesis need to take. (i) *In situ* PANI polymerisation on the surface of nanomaterials of the other part of the composite. (ii) One-pot one-step polymerisation of PANI and its composite material—this is common for polymeric support material. (iii) The last one is the physical mixing of the two materials that are separately synthesised to create a composite [58].

(i) Structure tailored polymerisation—in situ. *In situ* polymerisation of composite materials is usually done on the surface of the already-made nanocomposite materials. The process of synthesis may follow any of the different types of polymerisations available. Where aniline monomers are activated to polymerise in presence of activator APS under controlled temperature conditions and other special focus on nanostructures techniques are used. That is emulsion, dopant-assisted polymerisation and soft template use. This process has the potential to produce structure tailor nanocomposites on the basis that the other nanomaterials added may already have their definite structure. The graphenes for example are well-structured nanomaterials such that forming PANI on their surfaces is like overlaying a sheet on them. Highly definite shapes [59].

(ii) Structure-tailored polymerisation—one-pot synthesis. In this polymerisation technique, what is involved in the addition of various chemicals to the very same reactor [58]. The addition can be sequential at timed intervals with each addition tackling one part of the synthesis ultimately the product is PANI composite of the other materials. Pina et al. reported a one-pot synthesis of Fe_3O_4-PANI nanocomposite for super magnetic property evaluation [60]. The synthesis involved the mixing of aniline, iron ferrofluids and hydrogen peroxide as the oxidant. The spherical-shaped composite for Fe_3O_4-PANI formed.

(iii) Physical mixing of the already formed materials. The technique is the simplest as the PANI and the other composite materials are prior made. Zong et al. reported the synthesis of PANI-magnetic graphene oxide composite for the removal of metal waste and phenol, by pouring already synthesised PANI in a plasma reactor at adjusted conditions [61].

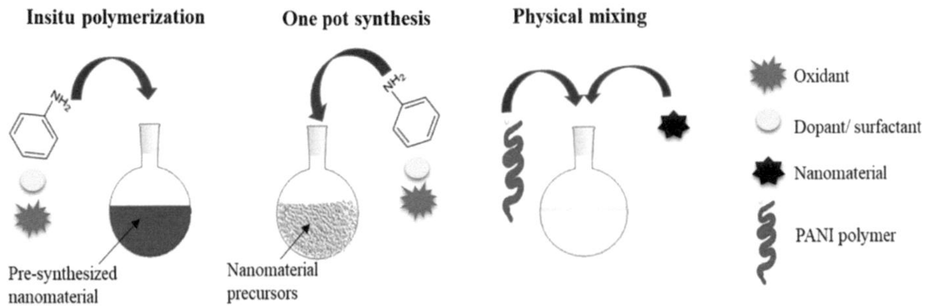

Figure 14.
Illustrates the PANI-composite synthesis techniques [58].

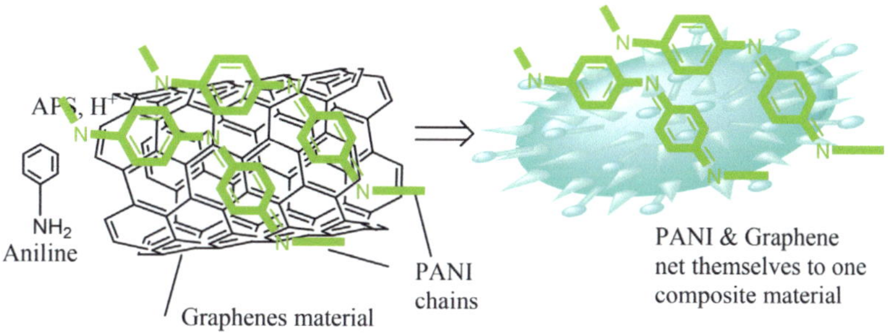

Figure 15.
Graphene-PANI-composite synthesised in situ general overview.

The PANI composite suprastructures are the diverse ones mostly for the creation of structured tailor PANI due to the many forms that can be manipulated. The summary of the synthesis could be illustrated as shown in **Figure 14**.

The composite molecule most applied is the graphene-based PANI material. This is because graphene by itself can be conducting material, while PANI too is conducting. The synergistic conduction becomes the alluring property the composite brings out. The general synthetic scheme follows the *in situ* polymerisation of the graphene and aniline monomer in one pot. The final product is the graphene around the PANI. **Figure 15** depicts the general reaction.

3. Conclusion

The synthesis techniques outlined above are the basic outline approach to conceptualise the process of designing the structure tailored PANI one desires. The actual reaction conditions and time may be adopted as per case analysis. However, every synthesis requires monomers, initiators and dopant material acids or redox-active chemicals. The actual synthesis for PANI with suprastructures that are tailored may need to follow many ways. Among the practical ways may involve the use of templates, both soft and hard ones. In either way, the template brings about the structural alignment of the monomers and subsequently the polymer formation follows the same aligned form. This in itself is beginning in the structured tailored polymerisation process. The other technique may involve the use of molecular imprinting. In this synthesis scheme for structure-tailored PANI, molecules to be imprinted are introduced in the polymer matrix to leave curvatures in a PANI, this is the highest form of tailor-made structures, but for suprastructures other forth mentioned techniques could still be used to bring about the structures of desire. In other techniques, dopants such as differently substituted sulfonic acids have shown great potential for synthesising the PANI of specific structures. While the molecules serve as dopants yet still work massively aligning molecules of PANI. The emulsion is another form of polymerisation that leads to the formation of PANI. Different shaped PANI from spheres, rods and fibres. Finally, the composites too are known to produce suprastructures that have tailored made structures. This is from the fact that by design the material to composite with may have already well-formed nature that acts as a frame for polyaniline to ride off as they form. In all the techniques mentioned above PANI is synthesised through the form of oxidation of aniline or anilinium ion.

Acknowledgements

The authors wish to acknowledge the fellow members of the faculty and department for moral support during the process of preparation. We also wish to acknowledge the Mulungushi University management for the financial support.

Conflict of interests

The authors declare they are no conflicts of interests.

Author details

Jimmy J. Daka* and George Mukupa
Department of Chemistry and Biology, Mulungushi University, Kabwe, Zambia

*Address all correspondence to: jdaka@mu.edu.zm

References

[1] Babel V, Hiran BL. A review on polyaniline composites: Synthesis, characterization, and applications. Polymer Composites. 2021;**42**:3142-3157. DOI: 10.1002/pc.26048

[2] Beygisangchin M, Abdul Rashid S, Shafie S, Sadrolhosseini AR, Lim HN. Preparations, properties, and applications of polyaniline and polyaniline thin films—A review. Polymers. 2021;**2021**(13):2003. DOI: 10.3390/polym13122003

[3] Syafei DI, Rini EP, Paristiowati M, Imaduddin A, Budi S. Synthesis and characterization of high conductivity polyaniline prepared at room temperature. Chemical Materials. 2022;**1**(1):7-11

[4] Tran BA, Duong HTL, Phan TXHTT. Synthesis and characterization of polyaniline–hydrotalcite–graphene oxide composite and application in polyurethane coating. RSC Advances. 2021;**2022**(11): 31572. DOI: 10.1039/d1ra04683g

[5] Majeed AH, Mohammed LA, Hammoodi OG, Sehgal S, Alheety MA, Saxena KK, et al. A review on polyaniline: Synthesis, properties, nanocomposites, and electrochemical applications. International Journal of Polymer Science. 2022;**2022**:19. DOI: 10.1155/2022/9047554

[6] Sai S, Kumar A. Synthesis and morphological study of polyaniline. European Journal of Molecular & Clinical Medicine. 2020;7(7):9

[7] Khalid NA, Razak JA, Hasib H, Ismail MM, Mohamad N, Junid R, et al. A short review on polyaniline (PANI) based nanocomposites for various applications: Enhancing the electrical conductivity. IOP Conference Series: Materials Science and Engineering. 2020;**957**:012028. DOI: 10.1088/1757-899X/957/1/012028

[8] Daka JJ, Munyati OM, Nyirenda J. Iron chlorophyll-a as biomimic catalyst for the green synthesis of polyaniline nanostructures: Evaluation, characterization and optimization. Sustainable Chemistry and Pharmacy. 2020;**15**:100194. DOI: 10.1016/j.scp.2019.100194

[9] Rahman S u, Röse P, Shah A u HA, Krewer U, Bilal S. An amazingly simple, fast and green synthesis route to polyaniline nanofibers for efficient energy storage. Polymers. 2020;**2020**(12):2212. DOI: 10.3390/polym12102212

[10] Munyati MO, Mbozi A, Siamwiza MN, Diale MM. Polyaniline nanoparticles for the selective recognition of aldrin: Synthesis, characterization, and adsorption properties. Synthetic Metals. 2017;**233**(2017):79-85. DOI: 10.1016/j.synthmet.2017.09.010

[11] Boeva ZA, Sergeyev VG. Polyaniline: Synthesis, properties, and application. Polymer Science. 2013;**56**(1):144-153. DOI: 10.1134/S1811238214010032

[12] Mazzeu MAC, Faria LK, Baldan MR, Rezende MC, Gonçalves ES. Influence of reaction time on the structure of polyaniline synthesized on a pre-pilot scale. Brazilian Journal of Chemical Engineering. 2018;**35**(1):123-130. DOI: 10.1590/0104-6632.20180351s1s201601

[13] Li B, Li Y, Ma P. Synthesis of different inorganic acids doped polyaniline materials and behavior of enhancing NH_3 gas sensing properties. Organic Electronics. 2023;**114**:106749. DOI: 10.1016/j.orgel.2023.106749

[14] Venugopal G, Quan X, Johnson GE, Houlihan FM, Chin E, Nalamasu O. Photo induced doping and photolithography of methyl-substituted polyaniline. Chemistry of Materials. 1995;**7**(2):271-276

[15] Malinauskas A. Self-doped polyanilines. Journal of Power Sources. 2004;**126**(1-2):214-220

[16] Bhadra S, Singha NK, Khastgir D. Dual functionality of PTSA as electrolyte and dopant in the electrochemical synthesis of polyaniline, and its effect on electrical properties. Polymer International. 2007;**56**(7):919-927

[17] Chiang J-C, MacDiarmid AG. Polyaniline: Protonic acid doping of the emeraldine form to the metallic regime. Synthetic Metals. 1986;**13** (1-3):193-205. DOI: 10.1016/0379-6779(86)90070-6

[18] Simon E. Ueber den flüssigen Storax (Styrax liquidas). Annals of Pharmaceuticals. 1839;**31**:265-277

[19] Rasmussen SC. The early history of Polyaniline II: Elucidation of structure and redox states. Substantia. 2021;**6**(1):107-119. DOI: 10.36253/Substantia-1425

[20] Pal R, Goyal SL, Rawal I, Gupta V. Dielectric characteristics of multiwall carbon nanotube-filled polyaniline. Materials Chemistry and Physics. 2023;**297**:127428. DOI: 10.1016/j.matchemphys.2023.127428

[21] MacDiarmid AG. A novel role for organic polymers (Nobel lecture). Angewandte Chemie International Edition. 2001;**40**:2581-2590

[22] Khalid M, Honorato AMB, Varela H. Polyaniline: Synthesis Methods, Doping and Conduction Mechanism. DOI: 10.5772/intechopen.79089

[23] Qiu B, Li Z, Wang X, Li X, Zhang J. Exploration on the microwave-assisted synthesis and formation mechanism of polyaniline nanostructures synthesized in different hydrochloric acid concentrations. Journal of Polymer Science Part A: Polymer Chemistry. 2017;**55**:3357-3369. DOI: 10.1002/pola.28707

[24] Wu W, Pan D, Li Y, Zhao G, Jing L, Chen S. Facile fabrication of polyaniline nanotubes using the self-assembly behavior based on the hydrogen bonding: A mechanistic study and application in high-performance electrochemical supercapacitor electrode. Electrochimica Acta. 2015;**152**:126-134. DOI: 10.1016/j.electacta.2014.11.130

[25] Stejskal J, Sapurina I, Trchová M. Polyaniline nanostructures and the role of aniline oligomers in their formation. Progress in Polymer Science. 2010;**35**(12):1420-1481. DOI: 10.1016/j.progpolymsci.2010.07.006

[26] Bhadra S, Khastgir D, Singha NK, Lee JH. Progress in preparation, processing and applications of polyaniline. Progress in Polymer Science. 2009;**34**(8):783-810. DOI: 10.1016/j.progpolymsci.2009.04.003

[27] Konyushenko EN, Stejskal J, Šeděnková I, Trchová M, Sapurina I, Cieslar M, et al. Polyaniline nanotubes: Conditions of formation. Polymer International. 2006;**55**:31-39. DOI: 10.1002/pi.1899

[28] Yitzchaik S. Enzyme mediated encapsulation of gold nanoparticles by polyaniline nanoshell. Journal of Self Assembly and Molecular Electronics. 2023;**3**(1):1-16. DOI: 10.13052/jsame2245-4551.311

[29] Jabłońska A, Gniadek M, Pałys B. Enhancement of direct electrocatalytic

activity of horseradish peroxidase on polyaniline nanotubes. Journal of Physical Chemistry C. 2015;**119**(22):12514-12522. DOI: 10.1021/acs.jpcc.5b03197

[30] Wang X, Schreuder-Gibson H, Downey M, Tripathy S, Samuelson L. Conductive fibers from enzymatically synthesized polyaniline. Synthetic Metals. 1999;**107**(2):117-121. DOI: 10.1016/S0379-6779(99)00150-2

[31] Cruz GJ, Morales J, Castillo-Ortega MM, Olayo R. Synthesis of polyaniline films by plasma polymerization. Synthetic Metals. 1997;**88**(3):213-218. DOI: 10.1016/S0379-6779(97)03853-8

[32] Kunicki L, Becker D, Fontana LC, Dalmolin C. Synthesis of polyaniline under electrolytic plasma on carbon fiber fabric. Brazilian Polymer Conference. 2020;**394**(10):2000018. DOI: 10.1002/masy.202000018

[33] Liu W, Yuan K, Qianxun R, Zuo S, Liang W, Yang S, et al. Functionalized halloysite template-assisted polyaniline synthesis high-efficiency iron/nitrogen-doped carbon nanotubes towards nonprecious ORR catalysts. Arabian Journal of Chemistry. 2020;**13**(4):4954-4965. DOI: 10.1016/j.arabjc.2020.01.018

[34] Wang J, Zhang S. Synthesis of polyaniline-sulfur composites with different nanostructures via an interfacial emulsification method and a micelle template method and their properties. RSC Advances. 2023;**10**(19):11455-11462. DOI: 10.1039/D0RA00122H

[35] Ren L, Li K, Chen X. Soft template method to synthesize polyaniline microtubes doped with methyl orange. Polymer Bulletin. 2009;**63**:15-21. DOI: 10.1007/s00289-009-0076-5

[36] Refaat D, Aggour MG, Farghali AA, Mahajan R, Wiklander JG, Nicholls IA, et al. Strategies for molecular imprinting and the evolution of MIP nanoparticles as plastic antibodies—Synthesis and applications. International Journal of Molecular Sciences. 2019;**20**(24):6304. DOI: 10.3390/ijms20246304

[37] Wackerlig J, Schirhagl R. Applications of molecularly imprinted polymer nanoparticles and their advances toward industrial use: A review. Analytical Chemistry. 2016;**88**:250-261. DOI: 10.1021/acs.analchem.5b03804

[38] Bagdžiūnas G. Theoretical design of molecularly imprinted polymers based on polyaniline and polypyrrole for detection of tryptophan. Molecular System Design Engineering. 2020;**2020**(5):1504-1512. DOI: 10.1039/D0ME00089B

[39] Tian X, Zhang B, Hou J, Gu M, Chen Y. In situ 2020 preparation and unique electrical behaviors of gold @ hollow polyaniline nanospheres through recovery of gold from simulated e-waste. Bulletin Chemical Society Japan. 2020;**93**(3):373-378. DOI: 10.1246/bcsj.20190286

[40] Pidenko PS et al. Molecularly imprinted polyaniline for detection of horseradish peroxidase. Analytical and Bio analytical Chemistry. 2020;**412**(24):6509-6517. DOI: 10.1007/s00216-020-02689-3

[41] Cao F, Liao J, Yang K, Bai P, Wei Q, Zhao C. Self-assembly molecularly imprinted nanofiber for 4-HA recognition. Analytical Letters. 2010;**43**(17):2790-2797. DOI: 10.1080/00032711003731480

[42] Kim SY, Song H-K. Conducting polymers with functional dopants and their applications in energy,

environmental technology, and nanotechnology. Clean Technology. 2015;**21**(1):12-21. DOI: 10.7464/KSCT.2015.21.1.012

[43] Hafizah MAE et al. Particle size reduction of polyaniline assisted by anionic emulsifier of sodium dodecyl sulphate (SDS) through emulsion polymerization. IOP Conference Series: Materials Science and Engineering. 2019;**515**:012080. DOI: 10.1088/1757-899X/515/1/012080

[44] Qiu B, Wang J, Li Z, Wang X, Li X. Influence of acidity and oxidant concentration on the nanostructures and electrochemical performance of polyaniline during fast microwave-assisted chemical polymerization. Polymers. 2020;**12**(2):310. DOI: 10.3390/polym12020310

[45] Ma Y, Zhang H, Hou C, Qiao M, Chen Y, Zhang H, et al. Multidimensional polyaniline structures from micellar templates. Journal of Material Science. 2017;**52**:2995-3002

[46] Wei C, Rakhi RB, Alshareef HN. Morphology-dependent enhancement of the pseudocapacitance of template-guided tunable polyaniline nanostructures. Journal of Physical Chemistry C. 2013;**117**(29):15009-15019. DOI: 10.1021/jp405300p

[47] Leng W, Zhou S, Guangxin G, Limin W. Wettability switching of SDS-doped polyaniline from hydrophobic to hydrophilic induced by alkaline/reduction reactions. Journal of Colloid and Interface Science. 2011;**369**(1):411-418. DOI: 10.1016/j.jcis.2011.11.080

[48] Zeng F, Qin Z, Liang B, Li T, Liu N, Zhu M. Polyaniline nanostructures tuning with oxidants in interfacial polymerization system. Progress in Natural Science: Materials International. 2015;**25**(5):512-519. DOI: 10.1016/j.pnsc.2015.10.002

[49] Jarai BM, Kolewe EL, Stillman ZS, Raman N, Fromen CA. Chapter 18 - Polymeric nanoparticles. In: Chung EJ, Leon L, Rinaldi C, editors. Micro and Nano Technologies, Nanoparticles for Biomedical Applications. Elsevier; 2020. pp. 303-324. DOI: 10.1016/B978-0-12-816662-8.00018-7

[50] Slomkowski S, Alemán JV, Gilbert RG, Hess M, Horie K, Jones RG, et al. Terminology of polymers and polymerization processes in dispersed systems (IUPAC Recommendations 2011). Pure and Applied Chemistry. 2011;**83**(12):2229-2259. DOI: 10.1351/PAC-REC-10-06-03

[51] Chajanovsky I, Suckeveriene R. Preparation of hybrid polyaniline/nanoparticle membranes for water treatment using an inverse emulsion polymerization technique under sonication. Processes. 2020;**8**:1503. DOI: 10.3390/pr8111503

[52] Österholm J-E, Cao Y, Klavetter F, Smith P. Emulsion polymerization of aniline. Synthetic Metals. 1993;**55**(2-3):1034-1039. DOI: 10.1016/0379-6779(93)90195-3

[53] Xavier Perrin F, Anh Phan T, Lam Nguyen D. Preparation and characterization of polyaniline in reversed micelles of decylphosphonic acid for active corrosion protection coatings. European Polymer Journal, Volume. 2015;**66**:253-265. DOI: 10.1016/j.eurpolymj.2015.01.052

[54] Ramanath P, Jeevananda T, Reddy KR, Raghu AV. Polyaniline-fly ash nanocomposites synthesized via emulsion polymerization: Physicochemical, thermal and dielectric properties. Materials Science for Energy

Technologies. Vol. 2021;**4**:107-112. DOI: 10.1016/j.mset.2021.02.001

[55] Rao PS, Sathyanarayana DN, Palaniappan S. Polymerization of aniline in an organic peroxide system by the inverted emulsion process. Macromolecules. 2002;**35**(13):4988-4996. DOI: 10.1021/ma0114638

[56] Vasiliev VV, Morozov EV. Chapter 1 - Introduction. In: Vasiliev VV, Morozov EV, editors. Advanced Mechanics of Composite Materials. Third ed. Elsevier; 2013. pp. 1-27. DOI: 10.1016/B978-0-08-098231-1.00001-7

[57] Ran F, Tan Y. Chapter 7 - Polyaniline-based composites and nanocomposites. In: Visakh PM, Pina CD, Falletta E, editors. Polyaniline Blends, Composites, and Nanocomposites. Elsevier; 2018. pp. 175-208. DOI: 10.1016/B978-0-12-809551-5.00007-2

[58] Hajjaoui H, Soufi A, Boumya W, Abdennouri M, Barka N. Polyaniline/nanomaterial composites for the removal of heavy metals by adsorption: A review. Journal of Composites Science. 2021;**5**(9):233. DOI: 10.3390/jcs5090233

[59] Guo H, Zhu H, Lin H, Zhang J. Synthesis of polyaniline/multi-walled carbon nanotube nanocomposites in water/oil microemulsion. Materials Letter. 2008;**62**:3919-3921. DOI: 10.1016/j.matlet.2008.04.016

[60] Pina CD, Rossi M, Ferretti AM, Ponti A, Lo Faro M, Falletta E. One-pot synthesis of polyaniline/Fe_3O_4 nanocomposites with magnetic and conductive behaviour. Catalytic effect of Fe_3O_4 nanoparticles. Synthetic Metals. 2012;**162**:2250-2258. DOI: 10.1016/j.synthmet.2012.10.023

[61] Zong P, Cheng Y, Wang S, Wang L. Simultaneous removal of Cd(II) and phenol pollutions through magnetic graphene oxide nanocomposites coated polyaniline using low temperature plasma technique. International Journal of Hydrogen Energy. 2020;**45**:20106-20119. DOI: 10.1016/j.ijhydene.2020.05.028

[62] Qiu H, Qi S, Wang D, Wang J, Xinming W. Synthesis of polyaniline nanostructures via soft template of sucrose octaacetate. Synthetic Metals. 2010;**160**(11-12):1179-1183. DOI: 10.1016/j.synthmet.2010.03.005

[63] Bhowmick B, Bain MK, Maity D, Bera NK, Mondal D, Mollick MMR, et al. Synthesis of dendritic polyaniline nanofibers by using soft template of sodium alginate. Journal of Applied Polymer Science. 2012;**123**:1630-1635. DOI: 10.1002/app.34575

[64] Djomehri S. Diffusive and Mechanical Properties of Biodegradable Alginate Stents. Master's Theses. 4230. 2012. 10.31979/etd.bkzn-84mp

[65] Razali NA et al. Preliminary screening oxidative degradation methyl orange using ozone/ persulfate. E3S Web of Conferences. 2018;**34**:02038. DOI: 10.1051/e3sconf/20183402038

[66] Hasnain MS, Jameel E, Mohanta B, Dhara AK, Alkahtani S, Nayak AK. Chapter 1 - Alginates: Sources, structure, and properties. In: Nayak AK, Hasnain MS, editors. Alginates in Drug Delivery. Academic Press; 2020. pp. 1-17. DOI: 10.1016/B978-0-12-817640-5.00001-7

[67] Kalaivasan N, Syed Shafi S. Synthesis of various polyaniline / clay nanocomposites derived from aniline and substituted aniline derivatives by mechanochemical intercalation method. Journal of Chemistry. 2010;7:7. DOI: 10.1155/2010/364680

[68] Conejo-Dávila AS, Moya-Quevedo MA, Chávez-Flores D, Vega-Rios A, Zaragoza-Contreras EA. Role of the anilinium ion on the selective polymerization of anilinium 2-acrylamide-2-methyl-1-propanesulfonat. Polymers. 2021;**13**(14):2349. DOI: 10.3390/polym13142349

[69] Palaniappan S, Anbalagan C. Polyaniline-dodecylhydrogensulfate-acid salt: Synthesis and characterization. Materials Chemistry and Physics. 2005;**92**:82-88. DOI: 10.1016/j.matchemphys.2004.12.033

[70] Liu W, Cholli AL, Nagarajan R, Kumar J, Tripathy S, Bruno FF, et al. The role of template in the enzymatic synthesis of conducting polyaniline. Journal of American Chemical Society. 1999;**121**(49):11345-11355. DOI: 10.1021/ja9926156

[71] Liu W, Kumar J, Tripathy S, Senecal KJ, Samuelson L. Enzymatically Synthesized Conducting Polyaniline. Journal of American Chemical Society. 2022;**121**(1):71-78. DOI: 10.1021/ja982270b

[72] Parthasarathy RV, Martin CR. Template-Synthesized Polyaniline Microtubules. Chemistry of Materials. 1994;**6**(10):1627-1632. DOI: 10.1021/cm00046a011

[73] Boeva Z, Sergeyev V. Polyaniline: Synthesis, properties, and application. Polymer Science Series C. 2014;**2014**:56. DOI: 10.1134/S1811238214010032

[74] Tran HD, Li D, Kaner RB. One-dimensional conducting polymer nanostructures: Bulk synthesis and applications. Advanced Materials. 2009;**21**:1487-1499. DOI: 10.1002/adma.200802289

[75] Liang H-W, Liu S, Yu S-H. Controlled synthesis of one-dimensional inorganic nanostructures using pre-existing one-dimensional nanostructures as templates. Advanced Materials. 2010;**22**:3925-3937. DOI: 10.1002/adma.20090439

[76] Shen Y, Qin Z, Li T, Chen Y, Liu N. Boosting the supercapacitor performance of polyaniline nanofibers through assisted oligomer assembly during seeding polymerization process. Electrochimia Acta. 2020;**356**:136841. DOI: 10.1016/j.electacta.2020.136841

[77] Goswani S, Nandy S, Fortunato E, Martins R. Polyaniline and its composites engineering: A class of multifuctional smart energy materials. Journal of Solid State Chemistry. 2023;**317**:123679. DOI: 10.1016/j.jssc.2022.123679

[78] Olad A, Ilghami F, Nosrati R. Surfactant-assisted synthesis of polyaniline nanofibres without shaking and stirring: Effect of conditions on morphology and conductivity. Chemical Papers. 2012;**66**:757-764. DOI: 10.2478/s11696-012-0197-4

Chapter 2

Polyaniline Derivatives and Their Applications

Hari Giri, Timothy J. Dowell, Mohammed Almtiri and Colleen N. Scott

Abstract

Polyaniline (PANI) is one of the oldest, yet most profound conducting polymer discovered. It's ease of synthesis, high conductivity, and environmental stability in the doped state makes it very attractive for a variety of potential applications. However, its insolubility and lack of redox stability has hindered many commercial applications. Consequently, many researchers have sought to overcome PANI's deficiencies in many ways including the development of PANI derivatives. This chapter will discuss the synthesis, properties, and applications of PANI derivatives. We will discuss three types of PANI derivatives—substitution on the benzene ring, substitution on the nitrogen atom, and fused ring cores. The properties of the PANI-derivatives will be compared to pristine PANI. Finally, we will emphasize the applications that arise from these derivatives and how they compare to PANI.

Keywords: Buchwald/Hartwig, polyaniline derivatives, conductive polymer, redox polymer, electroactive polymers

1. Introduction

Electronic devices based on inorganic conductors and semiconductors (e.g., Au, Cu, Pt, and Si) often require numerous etching and lithographic development steps during fabrication leading to longer processing times, costly equipment development, and higher priced devices [1]. In the middle of the twentieth century, these limitations spurred on extensive engineering and research efforts focused on optimizing fabrication methodologies; however, despite significant improvements, these devices are still limited by the physical properties of the inorganic materials they employ. Serendipitously, the discovery of *trans*-polyacetylene (PA) by Shirakawa [2–4] resulted in a collaboration with MacDiarmid and Heeger whose efforts led to the discovery of metallic-type electrical conductivity in halogen doped (i.e., *p*-doped) *trans*-polyacetylene in the late 1970s (**Figure 1a**) [4, 5]. In this way, the *p*-doping of PA (an electrical insulator with a non-zero energy gap (E_g)) induces uniform bond length for both the anti-bonding and bonding domains of the C$-$C (π^* and π orbitals, respectively). Doping also reduces the energy gap between PA's valance band (VB) and conducting band (CB) and increases charge carrier density and electrical conductivity (**Figure 1b**) [5]. Shortly after their discovery, numerous other organic polymers

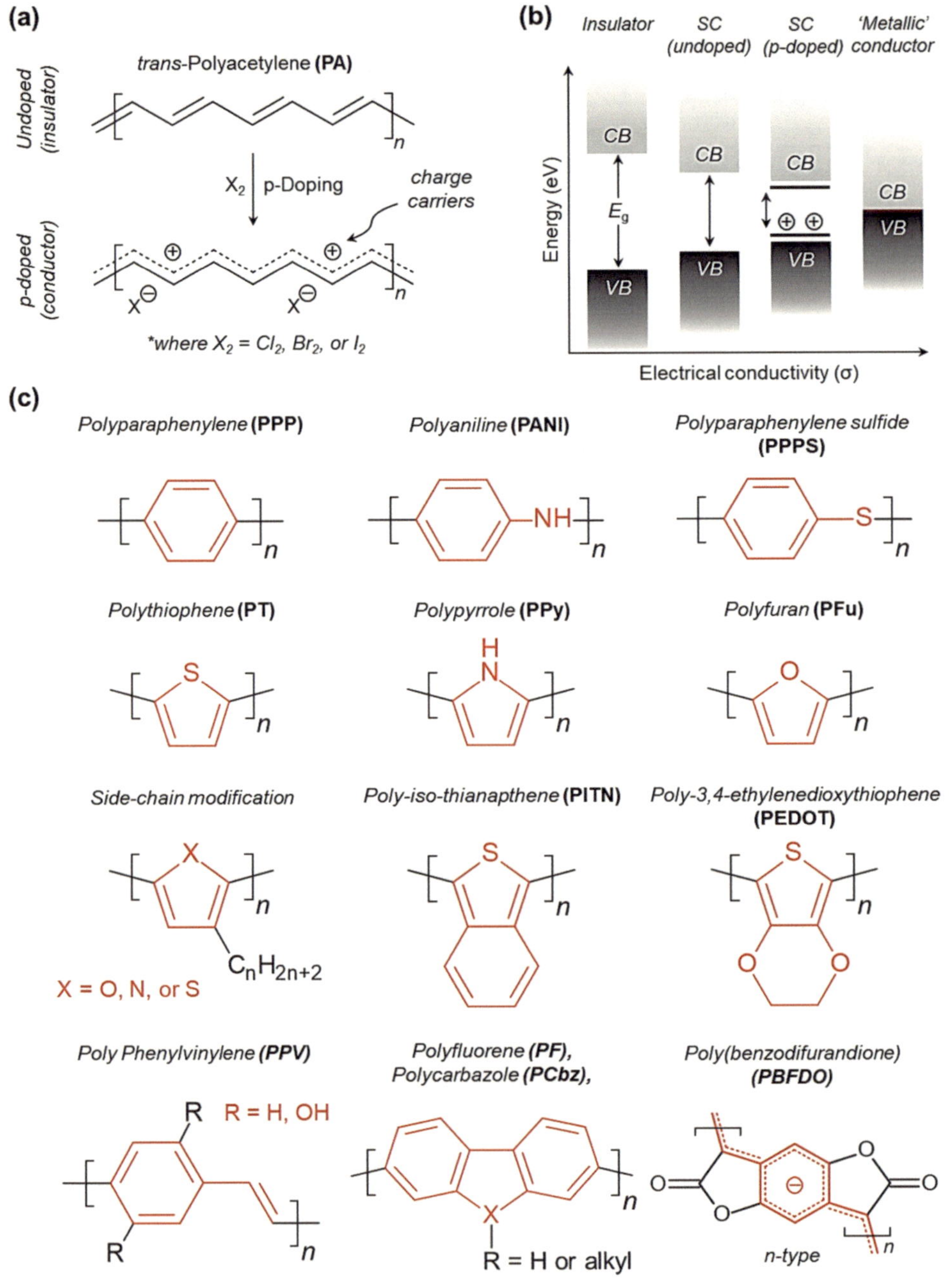

Figure 1.
(a) Illustration of the p-doping effect on PA, (b) various band gaps (E_g) of different material types and, (c) representative CPs and their chemical structures.

having similar characteristics where identified or "re-discovered" (e.g., polyaniline and polypyrrole); leading to a new family of materials aptly named conducting polymers (CPs) (**Figure 1c**) [6]. Similar to metals and semiconductors, CPs are characterized as being malleable and electronically conductive (with electrical conductivities ranging from semiconductive to metallic), as well as, having magnetic and optical properties similar to inorganic substances [5, 7]. Therefore, CPs are often referred to as "intrinsically conducting polymers" (ICP) or synthetic metals [1, 5, 8]. Similar to PA, ICPs have an innate inclination to conduct electricity since their structures contain a conjugated π-electron system, which forms uniform bond-lengths throughout the polymer backbone when doped [9].

The discovery of the properties of CPs has paved the way for creating new devices and novel applications where the benefits of plastics (i.e., scalability, processability, low-cost, etc.) and retention of high electrical conductivity are desired [10]. In this regard, solution and melt processable polymers could offer a facile approach to applying CP coatings and fabricating devices on large dimensional scales [11]. Furthermore, unlike their inorganic counterparts, CPs can potentially be processed using more modern solution-based processing techniques (e.g., inkjet printing), which could dramatically reduce the complexity, fabrication time, and cost when used to develop integrated electronic devices. Nevertheless, early development of CPs has lacked focus on developing truly "processable CPs" which has hindered their implementation to date.

A non-comprehensive list containing some classical CPs is shown in **Figure 1c**. Among them, polyaniline (PANI), polypyrrole (PPy), polythiophene (PT), and (poly-(3,4 ethylene dioxythiophene) (PEDOT) have been studied and employed to the greatest extent to date. Interestingly, although not fully appreciated at the time, the conductive nature of PANI was first observed more than 150 years ago after being prepared electrochemically from an acidic solution of aniline monomer [4, 6, 12]. Since then, PANI has become one of the most extensively studied CPs, especially over the last 20 years [12]. This interest in PANI is in large part due to its superior properties that impart high electrical conductivity, redox and ion-exchange properties, and high operational stability (i.e., is stable under a variety of environmental, thermal, and chemical conditions) (**Figure 2**) [13]. In addition, PANI's straightforward synthesis, tunable electrical conductivity, potential for chemical functionalization, and low synthetic cost make it an economically viable option compared to other CPs [8, 13–18]. PANI is therefore being investigated for use in various disciplines, including sensors [19], energy storage devices [20], corrosion inhibitors [21], solar cells [22], and electromagnetic shielding materials [23].

Doping of CPs can dramatically improve their electrical conductivity by many orders of magnitude and is required to obtain highly conductive materials. Doping usually occurs through an oxidation or reduction process and the polymers are considered p- and n-type, respectively [1, 7]. The oxidative/reductive doping process results in the alteration of the electronic configuration of the polymers making them unstable under environmental conditions. Another form of doping is protonic or acid doping. Unlike the oxidative/reductive doping, protonic doping (acid) does not alter the polymer's electron configuration; consequently, the polymer is environmentally stable in the doped state. It was shown that polyaniline could be doped with acid thus maintaining the same number of π-electrons in the backbone. Inorganic acids such as hydrochloric (HCl), sulfuric (H_2SO_4), nitric acid (HNO_3) and chloric acid ($HClO_3$) have traditionally been utilized as dopants in conductive PANI production. Polystyrene sulphonic acid (PSS), para-toluene sulfonic acid (*p*-TSA), camphor sulfonic acid (CSA), and dodecyl benzene sulfonic acid (DBSA) are all organic dopants that can also be used to dope PANI [24].

MacDiarmid proposed three forms for the structure of PANI based on the oxidation states [1, 25]. The leucoemeraldine base (LB) (colorless) is the completely reduced state of PANI, the emeraldine base (EB) (blue) is half-reduced and

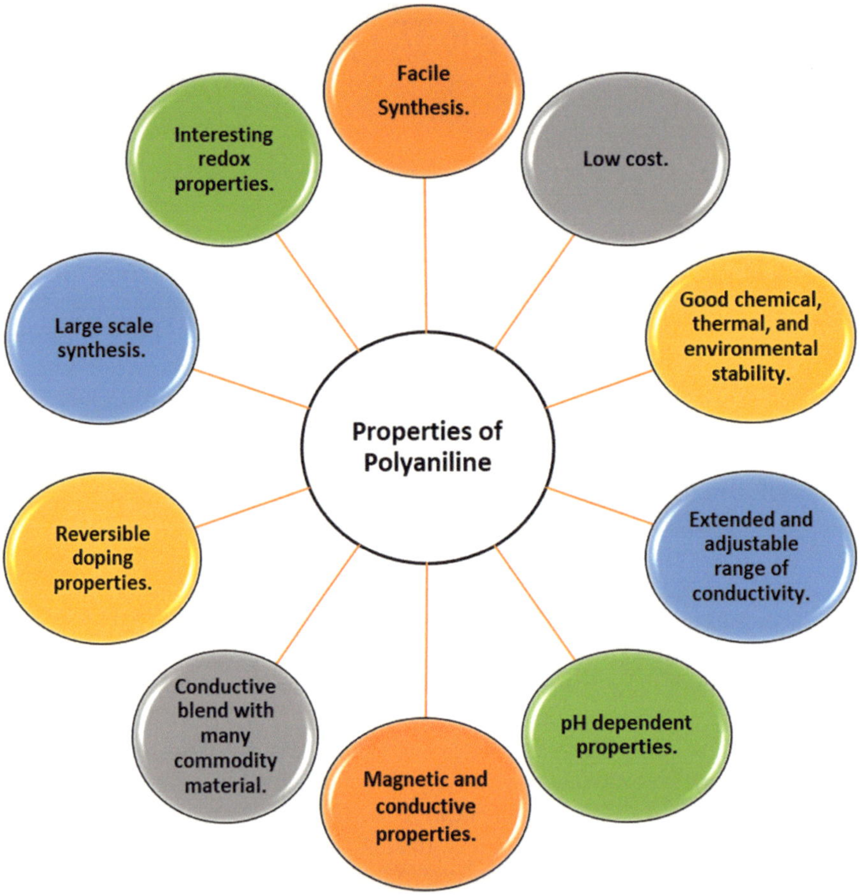

Figure 2.
Properties of polyaniline.

half-oxidized, and the pernigraniline base (PB) (blue/violet) is the fully oxidized form [1, 26, 27]. The leucoemeraldine base form has only benzenoid rings, the emeraldine base form has both benzenoid and quinonoid rings, while the pernigraniline base form has only the quinoid rings [18]. Upon doping, the emeraldine base is converted to the emeraldine salt (ES) (green) and becomes conducting. The transformation of the emeraldine base form of PANI from an insulating state to the conductive emeraldine salt form is due to the formation of polaronic carriers, i.e., the protonated species in the emeraldine state. The protonation occurs on the imine N rather than the amine N because imines are more basic than amines [24, 28, 29]. Consequently, protonic doping of the ES form produces a bipolaronic species (protonation of both imine Ns) that undergoes an internal relaxation to form the highly conductive polarons [29, 30]. Polaron separations results in charge flow. Due to its excellent stability at ambient temperature and the fact that after doping with acid, the resulting emeraldine salt form of polyaniline is highly electrically conductive, EB is regarded as the most practical form of polyaniline. Unfortunately, even after being doped with an acid, leucoemeraldine and pernigraniline still have low conductivity [1]. **Figure 3** shows the different oxidation states of PANI before and after protonic doping.

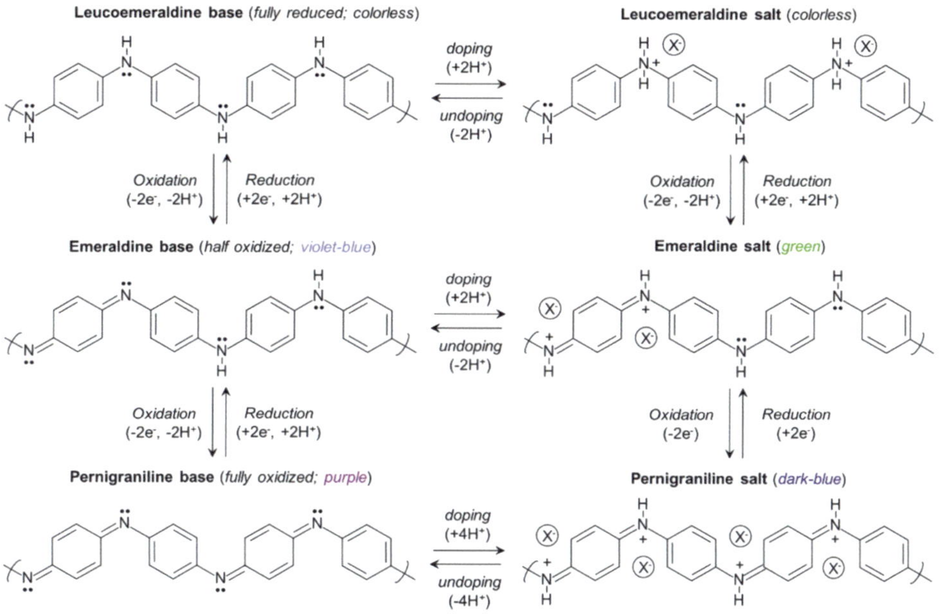

Figure 3.
Oxidation states and doping of various forms of polyaniline.

Despite the excellent properties and potential applications of PANI, it has some disadvantages that prevent its wide-spread applications. Actually PANI has been used in limited large scale commercial applications [31]. Of the major challenges to commercializing PANI, poor long term electrochemical stability and manufacturing costs due to low solubility in most solvents making processing difficult, are the two major issues [32]. The most successful improvements in PANI's processability have been achieved by using suitable dopants to increase the solubility and modifying the monomers to incorporate solubilizing alkyl chains. The stability of PANI is improved by forming composite materials [33–36]. In this chapter, we will focus on modifications of PANI (PANI derivatives) that address the limitations of poor solubility and electrochemical instability. The synthesis, characterization, and applications of these derivatives are the major focus here.

1.1 Synthesis of polyaniline

The three primary methods for producing PANI and its derivatives are chemical (i.e., oxidative polymerization), electrochemical (i.e., anodization), and more recently, palladium-catalyzed C-N cross-coupling (i.e., Buchwald-Hartwig reaction). Historically, chemical oxidative polymerization and electrochemical polymerization were the main methods of producing PANI due to their simplicity and availability of the monomers. Chemical oxidative polymerization involves an acidic aqueous media (pH < 2) with strong oxidants (e.g., potassium dichromate, hydrogen peroxide, ferric chloride, ceric nitrate, sulfate, and ammonium persulfate (APS)) in order to create the electrically conductive emeraldine salt form of PANI (**Figure 4**) [37, 38]. Unfortunately, the oxidative polymerization conditions produce PANI and PANI-analogs with variation in their structures and properties due to the many side reactions taking place simultaneously. While the overall mechanism is likely quite

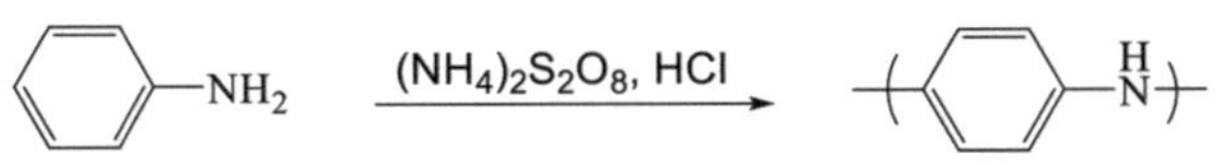

Figure 4.
Oxidative polymerization of aniline in acidic medium.

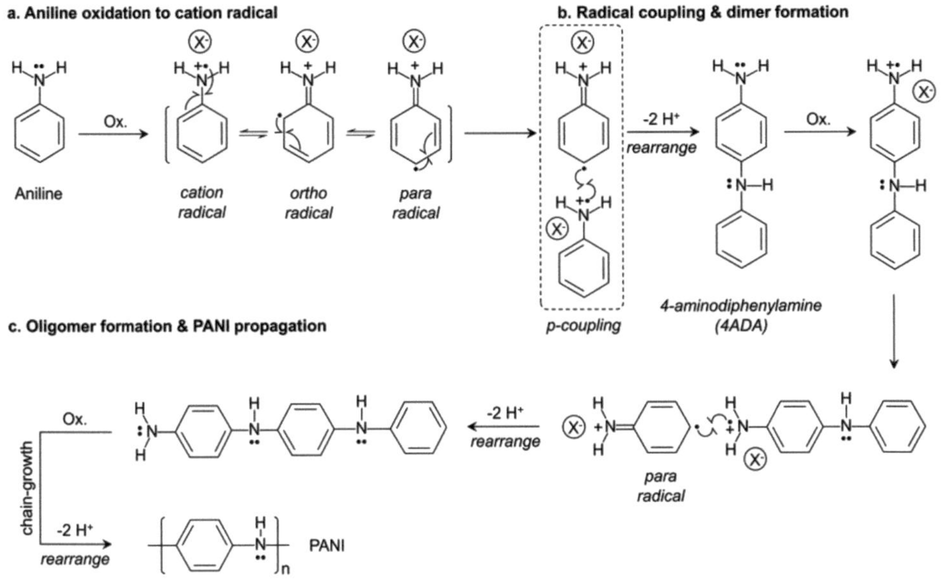

Figure 5.
Mechanism of polymerization for polyaniline. Adapted from Refs. [39, 40].

complex, a simplified mechanism for aniline polymerization in acidic media (pH < 2) is typically agreed to involve the initial formation of aniline radical cations (**Figure 5a**) that can undergo para-coupling forming a dimer (4ADA), due to the ortho/para (o, p) directing nature of the amine functional group. These dimers can likewise be oxidized to their cation radical form and couple with an additional aniline cation radical to form a trimer. This coupling process repeats until all monomer and/or oxidant expire, and the chain-growth process is terminated (**Figure 5b–d**) [13, 41]. The utility of this method enables easy scale-up to produce large quantities of PANI powders. As mentioned, the numerous side reactions and alternatively couplings likely occur simultaneously, which produces non-regiospecific PANI product [40, 42, 43]. However, PANI produced in this way does preferentially form p-substituted chains (despite equal likelihood for o-substitution), which has led to some authors suggesting an alternative initiation mechanism involving a phenazine initiation pathway (**Figure 6**) [39, 40].

Similar to the chemical-based oxidative polymerization route, electrochemical polymerization also involves the oxidation of aniline monomers, and its mechanism closely resembles that of oxidative polymerization (**Figure 5**) [39, 40]. Likewise, low pH (pH < 2) is preferred to achieve an electrically conductive PANI film. However, this process utilizes an applied potential at the electrode surface (e.g., using potentiostatic, galvanostatic, or cyclic voltammetry techniques) as opposed to an oxidizer in the bulk solution. In this way, the oxidation of aniline monomers to cation radicals occurs only at the electrode surface. These radicals then undergo coupling with one

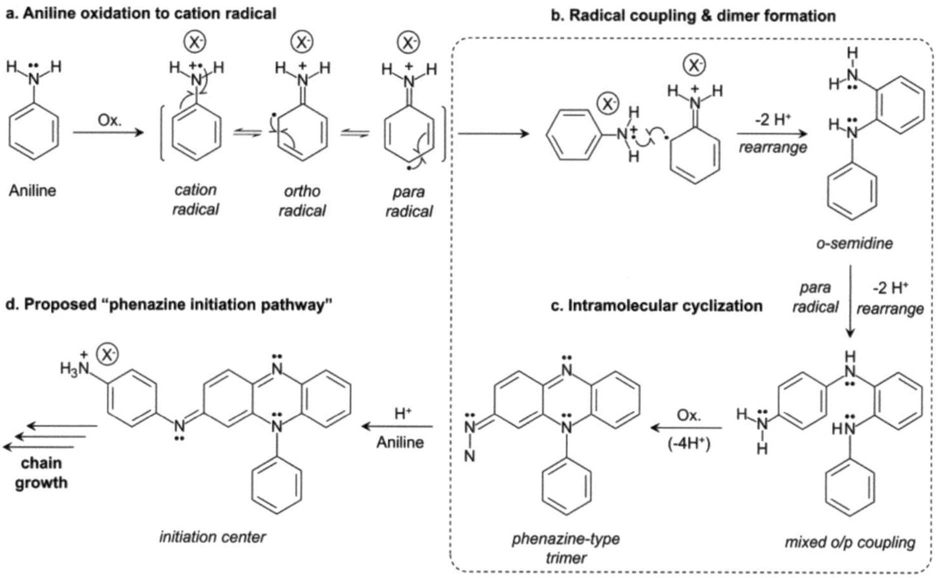

Figure 6.
Proposed alternative initiation pathway for polyaniline polymerization. Adapted from Ref. [39].

another to form dimers and ultimately polymers that becomes sufficiently large that it is no longer soluble in the electrolyte. Subsequently, the polymer forms a film at the electrode surface. The process can be terminated by removing the bias or exhausting the monomer [13, 39, 40]. This process typically produces insoluble and very high molecular weight PANI films with a fair amount of regio-irregularity (i.e., cross-linking).

Due to the presence of side reactions for both the chemical and electrochemical polymerization methods and the difficulty of obtaining a soluble PANI product from either methods, a new approach has recently been employed. This approach is based on the palladium-catalyzed C-N cross-coupling reaction developed by Buchwald and Hartwig that utilizes a Pd-catalyst, amine, aryl- halide, and hindered base in a suitable solvent [44–46]. The Buchwald/Hartwig reaction mechanism (**Figure 7a**) involves an oxidative insertion of the Pd(0) species I into an aryl halide substrate

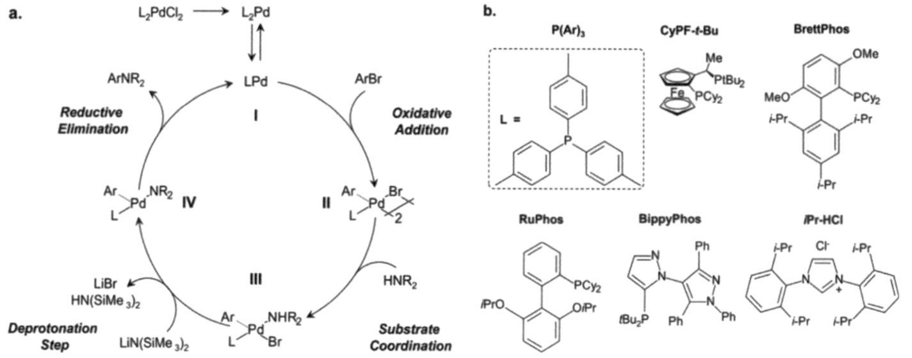

Figure 7.
Mechanism of the Buchwald-Hartwig palladium catalyzed C-N cross-coupling reaction and several optimized ligands.

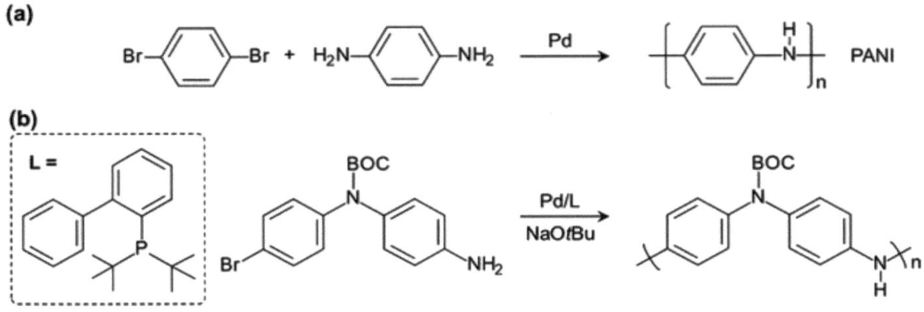

Figure 8.
PANI and a derivative by Buchwald/Hartwig reaction.

forming a Pd(II) species II. The increased acidity of the coordinated amine (III) at this point then allows for deprotonation with a hindered base (e.g., $HN(SIMe_3)_2$) and formation of a palladium-amido complex IV. Reductive elimination regenerates the Pd(0) catalyst I and produces an arylated-amine product. Significant improvements in ligand architecture (**Figure 7b**) and overall catalyst optimization have led to higher efficiency and broader application of the C-N coupling reaction. Consequently, coupling between numerous hetero(aryl) (pseudo)halides and amines at low temperature and low catalyst loading (ppm) is enabled. Buchwald and co-workers utilized this reaction to perform an efficient synthesis of a tert-butyloxycarbonyl (BOC) protected PANI (**Figure 8**) [47]. The BOC group, which can be removed during processing, was bulky enough to render the polymer soluble in common organic solvents. The Meyer group incorporated ortho-substituted repeat units into the backbone of PANI chain using the Buchwald/Hartwig reaction [48]. Due to the stepwise mechanism of the polymerization, alternative co-polymers can be produced. The *o,p*-PANI derivatives demonstrated similar properties (spectroscopic, oxidation states, colors, and conductivity) associated with the classical PANI. Furthermore, meta-substituted PANI can be prepared using this method [49–51], whereas the oxidative coupling or electrochemical reaction are unable to produce meta-substituted PANI.

2. Polyaniline derivatives

Although conductive polymers, particularly PANI, have numerous distinct benefits and possibilities as previously stated, PANI also has significant downsides. The fundamental disadvantage of PANI's conductive ES form is its low solubility in common organic solvents due to its stiff backbone, which makes it difficult to process. This insolubility hampers the practical application of PANI. Over the last few years, a lot has been done to make it easier to process PANI [11, 26, 52, 53]. Another, less studied problem of PANI is its electrochemical instability, which is mainly due to the defects in the polymer during the synthesis that resulted from the inherent instability of the radical species produced during the reaction [54–56]. The main approach to address the solubility, processability, and other aspects of PANI is to prepare PANI derivatives, which is discussed in this chapter. PANI derivatives can be broken down into three groups; (i) substitution on the benzene ring, (ii) substitution on the amino nitrogen, and (iii) fused ring systems (**Figure 9**) [26, 57].

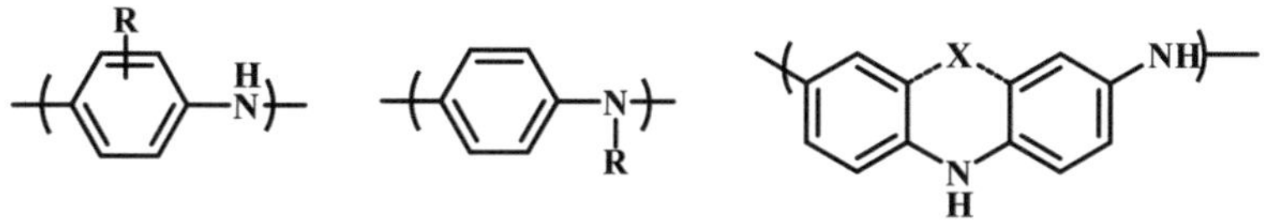

Figure 9.
Three types of PANI derivatives.

2.1 PANI derivatives with benzene ring substitution

Traditionally, increasing PANI's solubility was obtained by including substituents on the benzene rings, which weakens the chain's stiffness and interchain interaction [26, 58–61]. When substituents such as alkyl, alkoxy, and amino groups are attached to the aromatic rings, the steric barrier between the polymeric chains of the polymer increases, reducing the intermolecular force and increasing the dissolution of the polymer (**Figure 10**). For example, poly(2,3-dimethylaniline) (P(2,3-DMA)) was synthesized by Ma et al. using oxidative polymerization and was shown to have improved anticorrosion, electrochemical, and crystal properties compared to PANI [62, 63]. Unfortunately, no information was given by the authors pertaining to the solubility, conductivity, or electrochemical stability of the pristine polymer. Patil et al. prepared poly(o-anisidine) (POA) as coatings on copper by the electrochemical polymerization method [64]. Interestingly, the polymer film contained a mixed phase of both the pernigraniline base (PB) and emeraldine salt (ES) form of POA according to the optical absorption spectroscopy data. This polymer also lowered the corrosion rate of copper metal ~100 times more compared to the uncoated copper metal. However, due to the nature of the synthetic method that formed the film on the copper surface, there is no information about the solubility of the polymer. Sathyanarayana et al. prepared an ortho substituted poly ethoxy aniline (POEA) with excellent solubility and corrosion inhibition of iron in the presence of acid chloride [65]. A co-polymer, poly(o–/m-toluidine-co-o-nitroaniline), prepared by emulsion polymerization under oxidative conditions was also reported by the same authors. Interestingly, while the homopolymer of *o*-nitroaniline was not obtainable under the reaction conditions, the co-polymer with *o*-/*m*-toluidine occurred in low to moderate yields. The co-polymers showed improved conductivity, solubility, and thermal stability compared to the homopolymer of *o*-/*m*-toluidine [66].

To achieve water solubility of PANI, derivatives were prepared with water solubilizing hydrophilic groups such as a sulfonate [67, 68], carboxylate [69, 70], phosphonate, boronate [71], and hydroxyl groups) to the aromatic backbone (**Figure 11**). These groups preform dual function, i.e., they provide water solubility

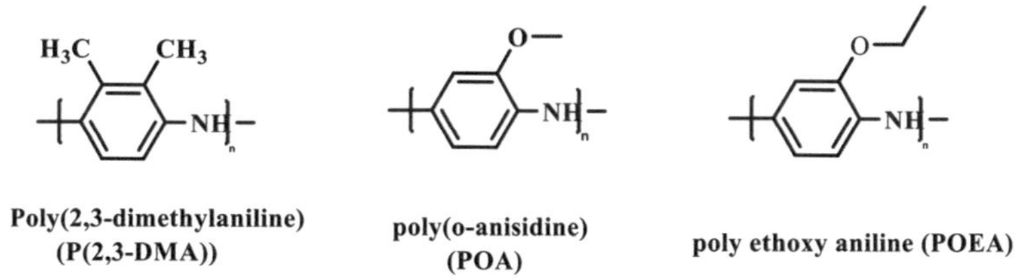

Figure 10.
Examples aryl substituted PANI derivatives.

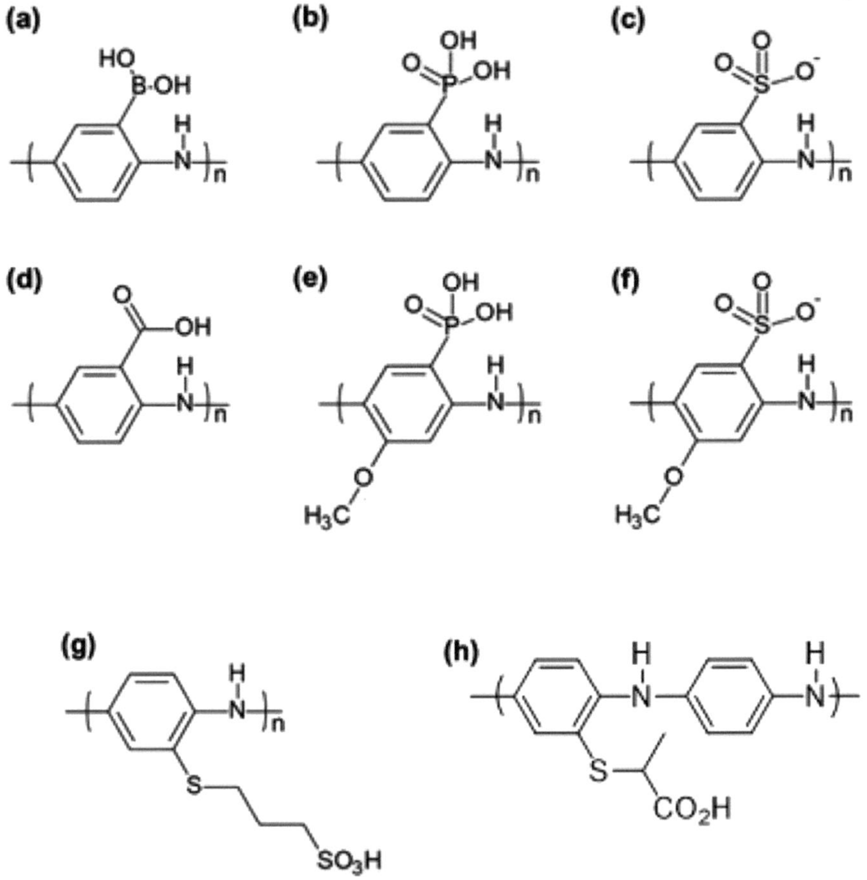

Figure 11.
Hydrophilic self-doped substituted derivatives of PANI.

and self-doping of the polymer. Sulfonic acid ring-substituted polyaniline (SPAN), first reported by Epstein and coworkers, was obtained by the sulfonation of the emeraldine base with fuming sulfuric acid [68, 72]. The transport, magnetic, and electrochemical properties of SPAN was later determined through a collaboration with the MacDiarmid group that provided insight to the effect of the sulfonate group on the solubility, doping mechanism, and charge transport of SPAN [73]. The polymer was shown to be self-doping as there was no pH effect on the conductivity for $pH < 7.5$. Interestingly, the polymer can be casted into a film from aqueous alkaline solvents (0.1 M NH_4OH) forming the blue-violet solution, which when casted onto a substrate and dried, reverts to the shiny green ES form that is highly conductive (0.1 S/cm).

Carboxylate substituted PANI were synthesized from anthranilic acid (*o*-aminobenzoic acid) and *m*-aminobenzoic acid by electropolymerization [70]. The sulphonated PANI, when compared to the carboxylate substituted PANI, was more electrochemically stable in the self-doped state than either the *m* or *o*-substituted carboxylate PANI. Another synthesis of copolymers from aniline and 2-substituted anilines with carboxyl groups (anthranilic acid, methyl anthranilate, and ethyl anthranilate) were described by Sá and co-workers [67]. The chemical oxidative method was used to prepare the polymers with a variety of acid dopants. Although the polymers contained variable amounts of phenazine-like by-products, the anthranilic ester polymers had high to moderate solubility in a variety of solvents and were more soluble than the other polymers. Moreover, the conductivities correlated well with the solubility and the methyl anthranilic ester polymer doped with HCl (PANMEA-HCl) gave the highest conductivity, which is comparable to PANI. These results lead to a conclusion that these polymers could be used for a variety of application that are relevant to PANI. A recent example revealed an interested thio-substituted lactic acid (TLA) PANI derivative that was conjugated to the benzene ring of PANI through a thiol bond [69]. The PANI derivative (PANI-TLA) was shown to be 4 orders of magnitude more conductive in alkaline media compared to PANI-base.

Some other interesting PANI derivative structures containing fused aryl and benzoquinone rings at the 2,3-positions of the PANI aryl group, have also been prepared. Poly(1-naphthylamine) [74, 75], poly(1-amino anthracene) [76–78], and copolymers from aminoquinoline(AQ) and anisidine (AS) [79] were prepared by both oxidative polymerization and electrochemical polymerization. The polymers have PANI-like features. They exist in several different oxidation states with conductivities ranging from 0.055–0.083 S/cm for the naphthalene and anthracene analogs. The conductivity was much less for the AQ/AS copolymer. PANI derivatives containing benzoquinone units fused to the 2,3-positions on the aniline moiety were also developed and their electrochemical properties with respect to cathode electrodes were studied. Some examples are shown in **Figure 12**. Poly(1,5-diaminoanthraquinone) [poly(DAAQ)] were prepared from 1,5-diaminoanthraquinone (DAAQ) by electrochemical polymerization and the conductivity values at different potential was measured [80]. It was shown that the conductivity values of the polymer ranged from 0.3 to 2.0 S/cm when the voltage range from −2.0 to 0.8 V, which are much higher than PANI at the lower potential. Poly(5-amino-1,4-naphthoquinone) (PANQ), reported by Pham et al., was prepared by electrochemical polymerization [81, 82]. The films had properties similar to PANI and the structure contained one quinone group per ANQ moiety. The polymer had a conductivity of 10^{-1} S/cm. Poly(5-amino-1,4-dyhydroxy anthraquinone) (PADAQ) was synthesized by oxidative polymerization for cathode material and was determined to behave like PANI [83]. Likewise, poly(1-aminoanthraquinone)

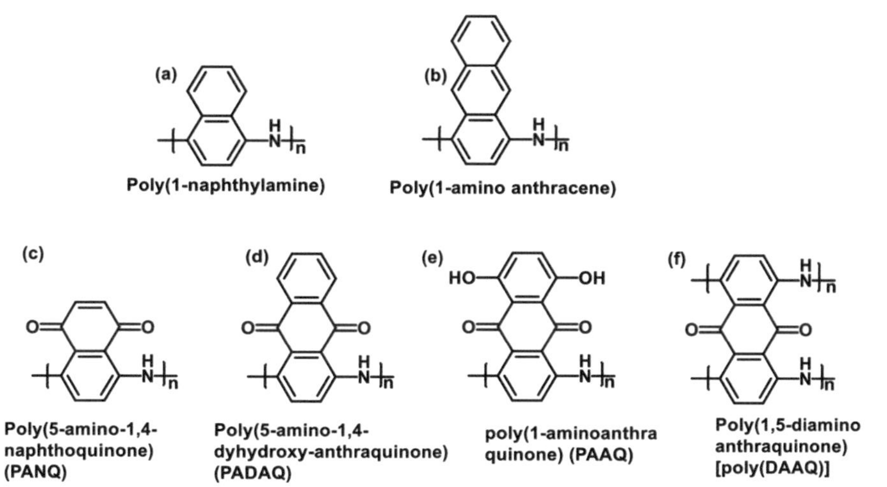

Figure 12.
Fused-aryl substituted PANI derivatives.

(PAAQ) films were prepared by the electrochemical polymerization and tested as efficient cathode material [84]. These polymers were studied for energy storage properties, which is discussed later.

2.2 Substituted derivatives of PANI on amino nitrogen atom

Another approach for increasing the solubility of PANI is to prepare N-substituted PANI derivatives (**Figure 13**) [85, 86]. While N-substitution has been demonstrated to have little effect on the synthesis of the polymers, it has significant effect on the properties of the final polymer due to the steric component of the substituent. The substitutions ranged from phenyl, substituted phenyl, to substituted alkyl chains. PANI N-substituted derivatives have improved solubility and processability; however, the final electronic properties are usually altered owing to the addition of the substituent on the N atom. One of the first studies of N-substituted PANI was reported by Chen and co-workers who prepared a series of N-alkylated PANI, where the alkyl chains ranged from butyl (C4) to the hexadecyl (C16) [86]. The solubility of the PANI derivatives in common organic solvents (tetrahydrofuran (THF), dichloromethane (DCM), and chloroform ($CHCl_3$)) increased with the size of the alkyl chain, especially above C-6. Lowering the polarity and stiffness of the polymer chains through integration of the flexible alkyl substituent is responsible for the improved solubility. On the other hand, N-alkylated EB become insoluble in highly polar solvents like N-methyl-2-pyrrolidone (NMP) and dimethyl sulfoxide (DMSO) due to lack the amine hydrogen necessary for hydrogen bonding. Importantly, the polymer existed in the EB form and can be effectively doped to produced soluble ES. The conductivity ranged from 10^{-4} to 10^{-2} S/cm with the C-6 alkyl chain having the highest conductivity. Other examples are poly-(aniline-co-N-(4-sulfophenyl)aniline) (PAPSA) [87] and its homo polymer, poly-(N-(4-sulfophenyl)aniline) [88], which were synthesized from aniline and sodium diphenylamine-4-sulfonate by the chemical oxidative polymerization. The polymer is soluble in aqueous basic media. While the conductivity is 10^2–10^6 times higher than other N-alkylsulfonate substituted PANI [89], it is 3 orders of magnitude lower than PANI. In general, N-substitution significantly lowers the conductivity of the polymer.

2.3 Fused-ring PANI derivatives

Fused-ring PANI derivatives have been prepared and studied since the 1960s [90]. These rigid ladder-type PANI derivatives are based on phenazine,

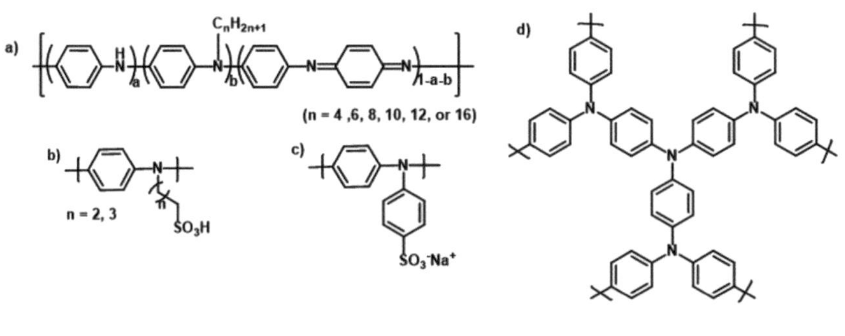

Figure 13.
N-substituted PANI derivatives.

phenoxazine, phenothiazine, and their co-polymers (**Figure 14**). The PANI derivatives maintain the PANI backbone with π-extended structures, which leads to mostly insoluble polymers [91, 92]. Fors and coworkers prepared PANI derivative poly(N-methylphenothiazine dimethylphenylenediamine) (PT-DMPD), from N-methylphenothiazine and dimethylphenylenediamine (**Figure 14a**), and poly(N-methylphenothiazine benzidine) (PT-BZ) from N-methylphenothiazine and tetramethylbenzidine using Buchwald/Hartwig reaction as redox active materials [93, 94]. Our group also synthesized PANI derivatives containing phenoxazine, carbazole, and phenothiazine heterocycles by the Buchwald/Hartwig cross coupling reaction to study their processability and electrochemical stability (**Figure 14b–d**) [95–97]. Unlike classical PANI, these polymers were electrochemically quite stable over 100 CV cycles without noticeable degradation while maintaining processibility by attaching long alkyl chain pendant. Overall, these PANI-analogs offer a unique insight into alternative ways of tuning PANI's structural, chemical, and physical properties. The electrical conductivities of the undoped (pristine) structures ranges from 10^{-5} to 10^{-6} S/cm and increase 2–4 orders of magnitude upon doping. More recent work has described additional side chain modification (**Figure 14b** and **d**), which can impart greater solubility for processable polymers. Orlov et al. describe the chemical oxidative polymerization of 2,5-dianiline-3,6-dichloro-1,4-benzoquinone (DADCB) to give a PANI derivative (Poly-DADCB) containing an N-substituted 5-aniline-3,6-dichloro-1,4-benzoquinone [98]. It was discovered that Poly-DADCB forms at a faster rate that PANI. However, the electrical conductivity was on the same order of magnitude as PANI.

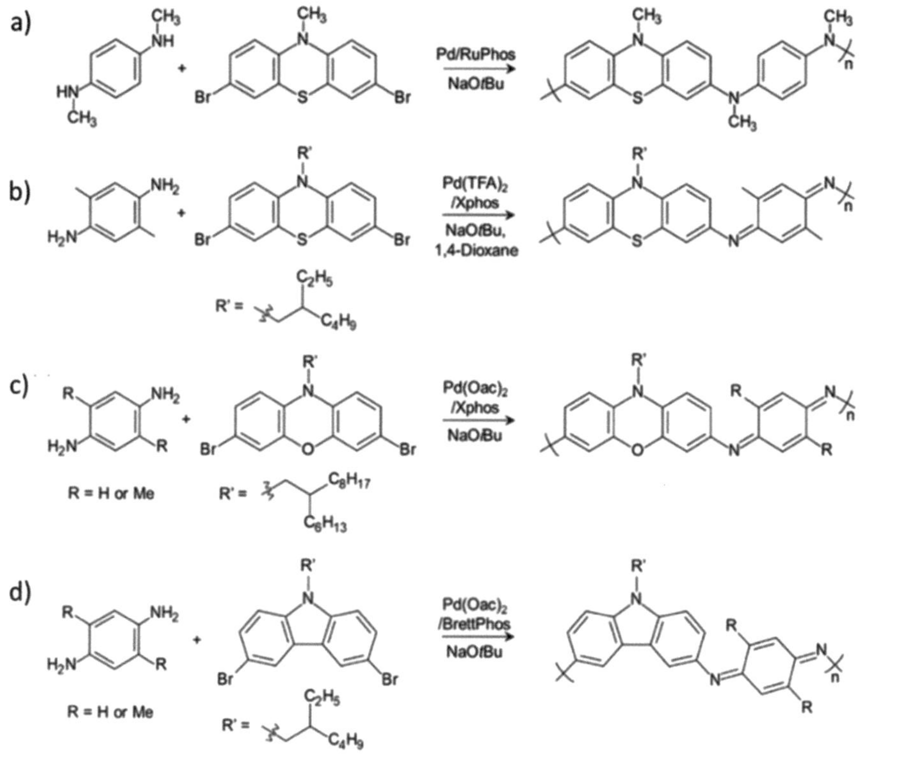

Figure 14.
Fused ring PANI derivatives.

3. Applications of PANI derivatives

PANI has been explored for many different types of applications such as biological sensors, anti-static wrappings, and electronic devices. Unfortunately, many of these applications are limited by its insolubility and electrochemical instability. However, PANI derivatives have been successful in achieving adequate solubility and electrochemical stability that they have become preferred to PANI in these applications. This section focuses on the applications of PANI derivatives in a variety of fields.

3.1 Antimicrobial activity

PANI and its derivatives have been shown to have antimicrobial properties where the mode of action is different for each derivative. For example, Andriianova et al. demonstrated the substituents effect of PANI derivatives on bactericidal properties of the polymer [99]. It was shown that PANI derivatives with substituents on the benzene ring, regardless of the bulkiness of the substituents, (POT, POA, PPA, PMA, and PClMA) had a negative or no effect on antibacterial activity compared to PANI. Alternatively, N-substituted PANI derivatives (PNClA and PNMA) increased the antibacterial properties against gram-positive (*Bacillus subtilis*) and gram-negative (*Pseudomonas aureofaciens*) bacteria (**Figure 15**). Additionally, these N-substituted.

PANI derivatives also showed bacteriostatic properties towards the microorganisms. Here it was determined that the mode of bactericidal activity is due to the positive charges on PANI and its derivatives that disrupt the nature of the bacteria's cell membrane that ultimately leads to cell lysis. In another example, PANI and poly(3-aminobenzoic acid) (P3ABA) (**Figure 16a** and **b**) were investigated as antimicrobial agents and found that the mechanism of action is different for each polymer. While both polymers adversely affected the growth of bacteria, the mechanism of antimicrobial activity for PANI is believed to be a result of the production of hydrogen peroxide (H_2O_2) that leads to hydroxyl radicals. On the other hand, the antimicrobial activity of P3ABA is due to the disruption of the metabolic and respiratory machinery of the bacteria by targeting the ATP synthase, thus causing acid stress [100]. PANI-based materials have also been investigated in tissue engineering applications,

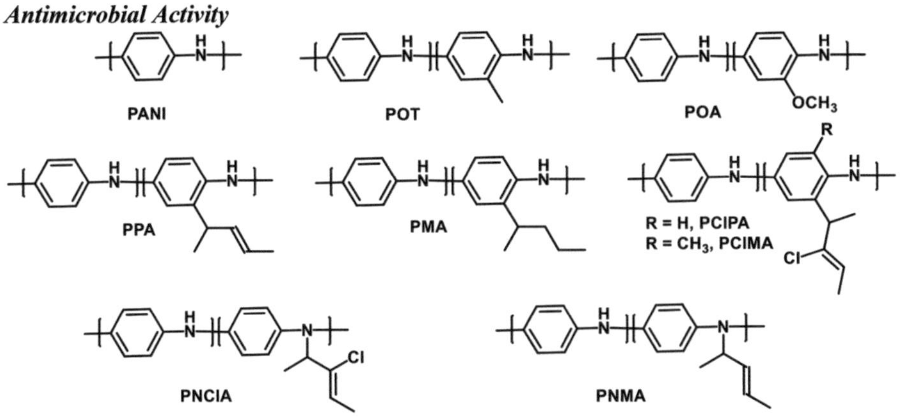

Figure 15.
Structures of PANI-derivatives with antimicrobial activity. Structures drawn from Ref. [99].

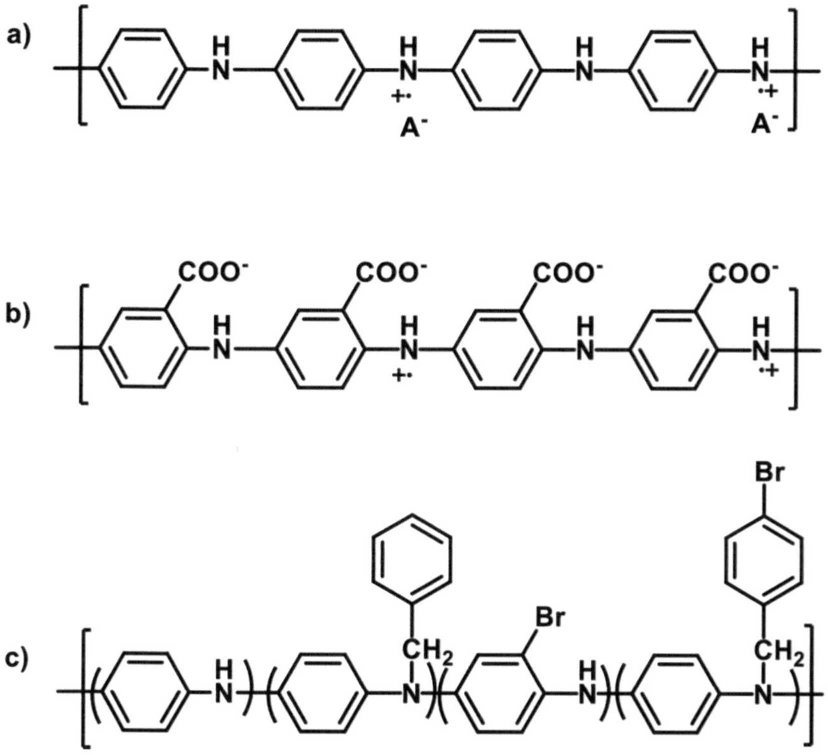

Figure 16.
(a) PANI (b) P3ABA, and (c) BBP. Structures drawn from Refs. [100, 101].

which was recently reviewed [102]. Benzyl-substituted PANI (BP), bromine-benzyl-disubstituted PANI (BBP), (**Figure 16c**) and PANI nanoparticles were tested against both Gram-negative E. *coil* and Gram-positive *B. subtilis*; however, it was sown that only BBP had high antibacterial activity where as PANI and BP did not [101].

3.2 Bio- and chemo-sensors

3.2.1 Biosensors

PANI and its derivatives are explored as biosensors. PANI derivative with Au/Pd composites were prepared and analyzed as biosensors for cancer biomarkers [103]. Specifically, the polymers, poly(N-methyl-*o*-benzenediamine), poly(N-phenyl-*o*-phenylenediamine), poly(N-phenyl-*p*-phenylenediamine), and poly(3,3′,5,5′-tetramethylbenzidine) all with Au/Pd nanocomposite were used to simultaneously detect four tumor biomarkers, carcinoembryonic antigen (CEA), carbohydrate antigen 19-9 (CA199), carbohydrate antigen 72-4 (CA724), and alpha fetoprotein (AFP) (**Figure 17**). Interestingly, these biomarkers can be found in different cancer cells and as such the PANI dervatives/Au/Pd nanocomposities are capable of detecting multiple cancers simultaneously. For example, biomarkers CEA, CA199, CA724 are found in gastric cancer where as liver cancer can be diagnosed from biomarkers CEA, CA199, and AFP. As such, these biosensors can be used to detect gastric and liver cancer simultaneously. In another example, polyaniline (PA), poly(o-anisidine)

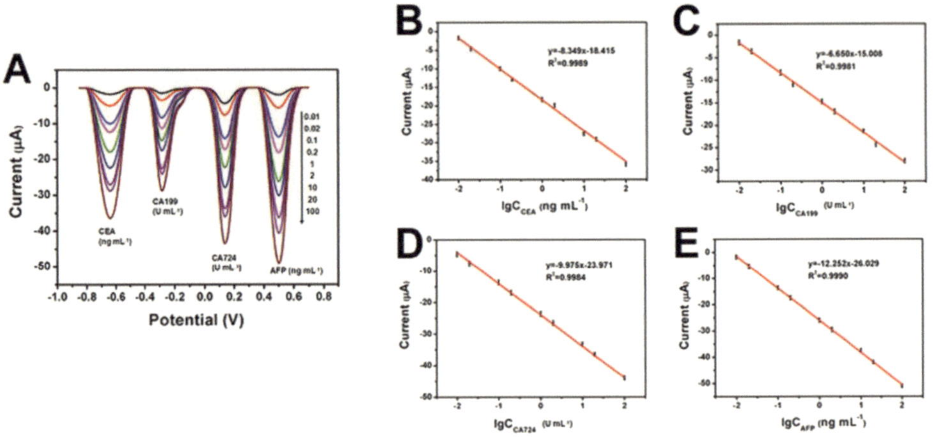

Figure 17.
SWV responses (A) and calibration curves for different concentration of CEA (B), CA199 (C), CA724 (D), and AFP (E) in PBS (pH 5.5) with 1.5 mM H_2O_2. Reproduced from Ref. [103].

(POA) and their co-polymer poly(aniline-co-o-anisidine) (PA-co-POA) thin films were used as glucose sensor by trapping glucose oxidase (GOD) in the polymers' films [104]. Amperometric measurements show that the PA GOD electrode responded the fastest to glucose followed by PA-co-POA, and the least responsive was POA GOD electrodes.

3.2.2 Chemosensors

PANI has also been widely explored as chemosensors due to its high conductivity and electrochromic properties. However, there are only few examples of PANI derivatives in sensor applications. PANI derivatives containing *ortho*-substituted methyl and methoxy groups, *ortho,meta*-disubstituted methyl, N-substituted methyl, ethyl, and phenyl groups were investigated for their sensory properties towards C-1 to C-5 alcohols [60]. The results revealed that resistance in the polymer increased (conductivity decreased) for all the polymers in the presence of C-1 to C-3 alcohols, but decreased (conductivity increased) for butanol and heptanol. A change in the crytallynity of the polymers due to the incorporation of alcohol between the chains is deemed responsible for the results. Consequently, the chain length of the alcohol and their structures can make a big difference in their interaction with the polymers' inter and intra chain packing. Poly(o-alkylanilines) such as poly[2-(1-methylbut-1-en-1-yl)aniline] (P-PA), poly[2-(1-methylbutyl)aniline] (P-MB), and poly[2-(2-aminophenyl)pentan- 2-ol] (P-AP) have also shown to have high sensitivity to moisture and ammonia (**Figure 18**) [19]. P-PA had the largest response to moisture, while P-MB was the most sensitive to ammonia. The results were explained based on the roughness of the films. For example, P-PA had the smoothest surface and gave the highest conductivity in the presence of moisture, while P-MB was the roughest film but had the best interaction with ammonia vapors. Recently, two-dimensional (2D) fluorescent PANI derivatives (PPED and PPBA) were used to demonstrate high selectivity and sensitivity in the detection of iron ion (Fe^{3+}), copper ion (Cu^{2+}), and hydrochloric acid (HCl) [105], while PANI derivatives from carbazole and fluorene were used as pH sensors in both solution and gas phase [106].

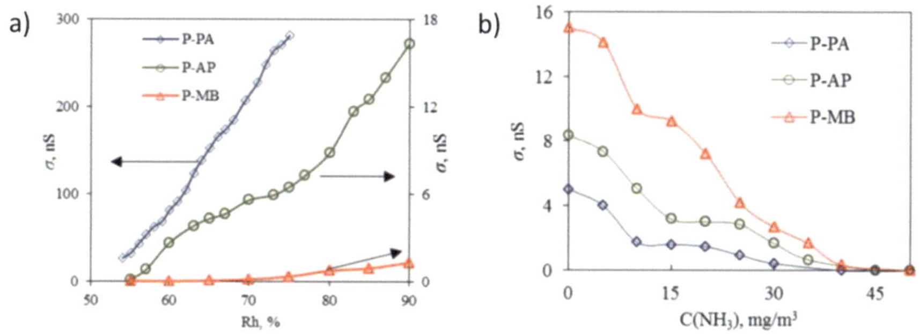

Figure 18.
Humidity and Gas sensors. (a) Plots of the conductivity of P-PA, P-AP and P-MB films on the air humidity. (b) Plots of the conductivity of P-PA, P-AP and P-MB films on the concentration of ammonia vapor. Reproduced from Ref. [14] with permission from the Royal Society of Chemistry.

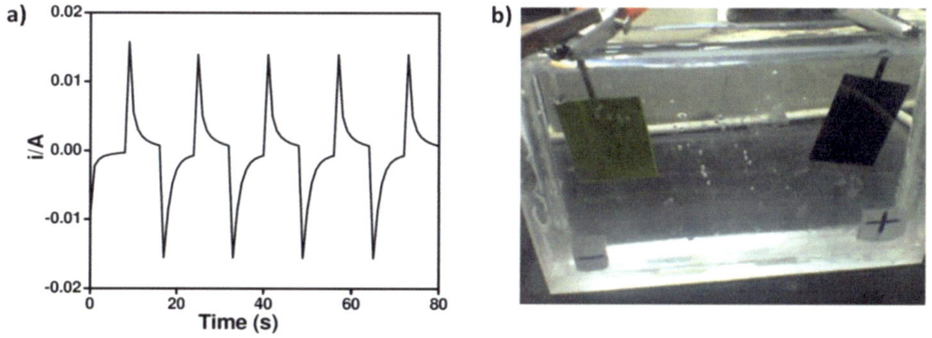

Figure 19.
Chronoamperometry of poly(OMA) and electrochromic devices of poly(OMA). Reproduced from Ref. [102] with permission under Creative Commons License.

3.3 Electronic devices: electrochromic glass, solar cells, and LEDs

PANI is well known for its electrochromic properties as the polymer changes color reversibly with an electric field [107]. These materials are usually desirable for smart glass. A similar behavior is also demonstrated for its derivatives. For example, poly(2-methoxyaniline) and its copolymer, poly(2-methoxyaniline-co-3-aminobenzene sulphonic acid), were shown to have electrochromic behavior in a display device by Saharan et al. [108]. The response time was determined to be about 6 s going between the oxidized and reduced states (**Figure 19**). Additionally, the cycling behavior of the device was deemed to be stable. A PANI derivative containing phenoxazine unit was shown by Almtiri et al. to undergo similar color changes based on the oxidation state of the polymer [96]. Between 0 and 0.4 V, the polymer is light yellow, which changes to dark green between 0.4 and 0.6 V. Further oxidation renders the film deep blue at voltages greater than 0.6 V (**Figure 20**). The cycling stability was measured by cyclic voltammetry (CV) and was shown to be very stable over 100 cycles. Electrochromic conducting copolymers from aniline and different feed ratio of 4,4′-diaminodiphenyl sulfone (DDS) were deposited on ITO glass plates for spectrochemical analysis [109]. The copolymers displayed multiple colors similar to PANI when the applied potential was switched between −0.2 and 1.2 V, where the color changed from yellow to green

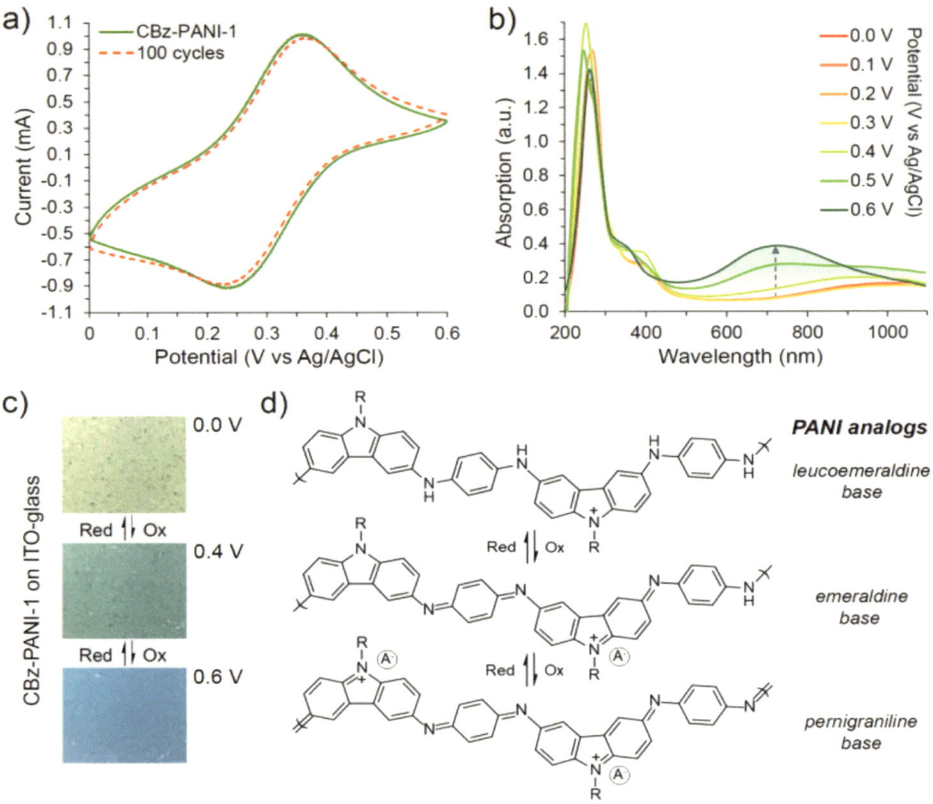

Figure 20.
Electrochemical and electrochromic behavior of carbazole PANI. Reprinted with permission from Ref. [96]. Copyright 2022 American Chemical Society.

and then blue. Upon changing the voltage from −0.2 to 0.0 V, a neutral yellow color formed. Further oxidation from 0.1 to 0.8 V, resulted in the conductive green color known to PANI and its derivatives. From 1.0 to 1.2 V, the polymer formed the blue color known to the fully oxidized pernigraniline form.

PANI and its derivatives are utilized as hole transport layers in Perovskite solar cells (PSC). For example, poly(*o*-methoxyaniline) (PoMA) doped with 4-dodecylbenzenesulfonic acid (DBSA) was used to replace spiro-OMeTAD, which was the most explored hole transport material (HTM) in PSC. The best power conversion efficiency (10.05%) was obtained using doped PANI-DBSA with Au as contact, which is comparable with the device using Spiro-OMeTAD. Unfortunately, the device of PoMA with silver contact had the best performance of 5.5%, which is significantly lower than the pristine PANI [110]. This outcome is probable due to the close alignment of PANI and perovskite's HOMO levels versus PoMA and perovskite's HOMO levels (**Figure 21**). Our group have also prepared three PANI derivatives, PANI–carbazole (P1), PANI–phenoxazine (P2), and PANI–phenothiazine (P3), with different energy levels that were used to form an interface between PEDOT:PSS and the perovskite layer [97]. While all the devices from all the polymers performed better than the device without them, P2 exhibited the highest V_{oc} of 1.13 V and power conversion efficiency of 21.06%. The result is also explained by the close overlap of the perovskite's HOMO to P2 versus the other polymers (**Figure 22**).

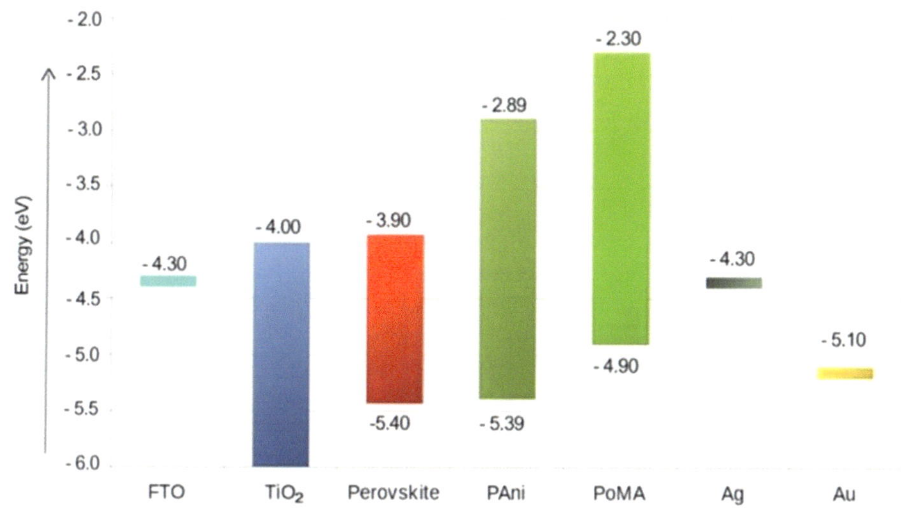

Figure 21.
Energy diagram of the different materials used in perovskite solar cells assembled in this work. Reprinted from Ref. [110].

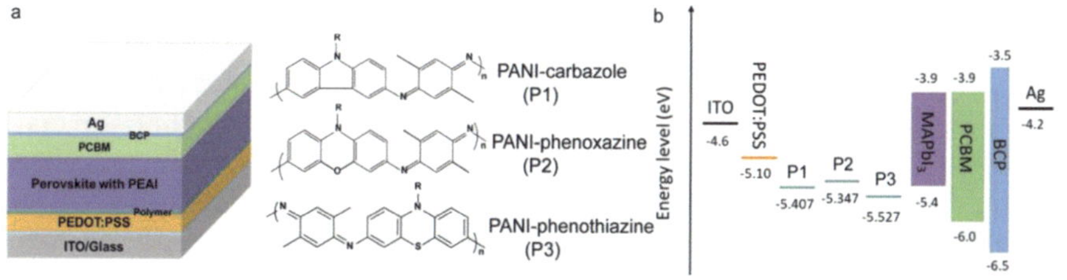

Figure 22.
(a) MAPbI3-based PSC device configuration and the chemical structure of three polymers. (b) Energy level alignment of the polymers in the MAPbI3-based PSC devices. Reproduced from Refs. [91, 97] with permission from John Wiley and Sons.

3.4 Anti-corrosion and anti-fouling

PANI and its derivatives have been widely explored as anti-corrosion materials [111]. Trimethylsilyl-substituted polyaniline (PSiAn) was synthesized and used as anticorrosion materials [112]. The fabricated filler was highly hydrophobic due to the trimethylsilyl group. It was incorporated into the epoxy matrix to provide a low porous material with high barrier to aggressive corrosive substances. Anticorrosion coatings prepared from poly (aniline-co-2-ethylaniline) (PEA) micro/nanostructures displayed good anticorrosion performance with corrosion protection efficiency of 87.29% [113]. The hydrophobic property (CA = 145°), low conductivity, and low porosity was attributed to the anticorrosion performance of the composite material. Chloro- and bromo-substituted PANI derivatives have been shown to have excellent anti-corrosion and anti-fouling properties. Aluminum alloy 3105 (AA3105) was coated with poly(2-chloroaniline) (PClAni) by electropolymerization and the polymer was investigated for its anticorrosion properties. Potentiodynamic polarization technique and electrochemical impedance spectroscopy data showed that the metal coated with PClAni had outstanding performance against corrosion versus the untreated metal when subjected to 3.5% NaCl solution [114]. A similar result was obtained with bromo-substituted PANI derivatives (Br-PANI). De-doped Br-PANI in epoxy resin with varying amounts of Br substituents (EBP coatings) were prepared and investigated for anti-fouling and anti-corrosion properties by the accelerated immersion test, electrochemical impedance spectroscopy (EIS), XPS, antibacterial test, and field test [115]. It was shown that the polymers possessed excellent anticorrosion properties after immersion in 12.0 wt% NaCl solution at 95°C for 100 days. Furthermore, the EBP coatings also demonstrated better antibacterial and antifouling performance when compared to pure epoxy coating or de-doped PANI composite coatings. Poly(1-naphthylamine) (PNA) composites with polyvinyl alcohol (PVA) and polyvinylchloride (PVC) have also shown both anticorrosion and antifouling properties. PNA/PVA coatings have shown good corrosion protective efficiency and resistance in acid, alkaline, and saline media [116]. The same polymer composite also demonstrated antibacterial activity against *Escherichia coli* and *Staphylococcus aureus* [117]. Examples of the structures for anti-corrosion and anti-fouling are shown if **Figure 23**.

3.5 Energy storage devices: supercapacitors and batteries

Supercapacitors are energy storage devices that are characteristically high in energy and power density, as evidenced by a more extended battery life and fast charge/discharge. They are the future of portable consumer electronics that have excellent operational lifetimes. Supercapacitor energy storage devices have 10 times more operational lifetimes than lithium-ion batteries. Some studies have focused on enhancing the capacity of the materials and PANI derivatives are among the materials being studied for increased capacitance. PANI derivative, poly(DAAQ), was investigated as an electrochemical capacitor. The constructed symmetrical (poly(DAAQ)/poly(DAAQ)) electrochemical capacitor was shown to exhibit high specific energy (25–46 Wh/kg) and high specific power (10,200–30,500 W/kg) at discharge rates from 30 to 90 C [80]. Our group also developed supercapacitor from PANI derivative containing the carbazole unit. Cbz-PANI was used to construct electrodes for a supercapacitor device that showed a maximum areal capacitance of 64.8 mF cm^{-2} and a specific capacitance of 319 F g^{-1} at a current density of 0.2 mA cm^{-2} in a symmetrical device (**Figure 24**) [96]. Moreover, the electrode showed excellent cyclic stability (≈ 83% of capacitance retention) over 1000 CV cycles and a 91% capacitance retention after 1000 cycles of charge and discharge in a device.

On the other hand, batteries provide high energy densities that can be delivered over time. Poly(5-amino-1,4-dyhydroxy anthraquinone) (PADAQ) was used as the cathode material in a lithium ion battery [83]. The initial discharge capacity was 101mAhg^{-1} at the current density of 400 mAhg^{-1} with the cutoff voltage of 1.5–3.7 V. At a high current density of 1400 mAhg^{-1}, the capacity of the polymer was 95 mAhg^{-1}. Poly(5-amino-1,4-naphthoquinone) (PANQ) recorded an experimental

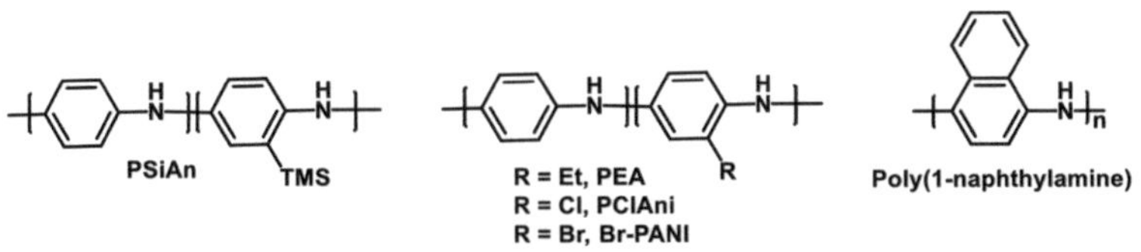

Figure 23.
Structures of anti-corrosion/anti-fouling PANI derivatives.

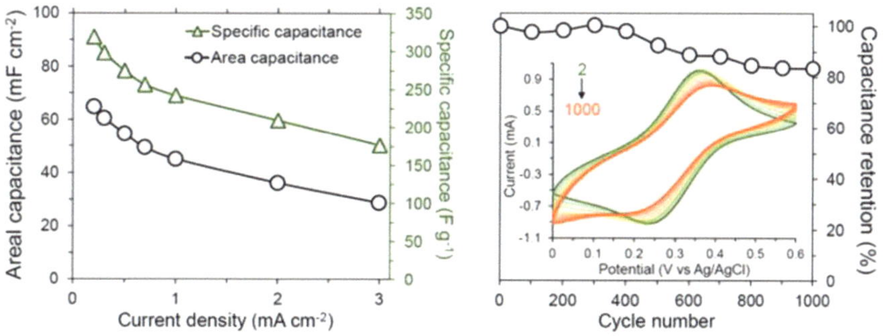

Figure 24.
Areal capacitance, specific capacitance, and capacitance retention of Cbz-PANI-1/AU@PET. Reprinted with permission from Ref. [96]. Copyright 2022 American Chemical Society.

charge storage capacity of 220–290 Ah/kg in nonaqueous electrolytes. The specific energy of the battery is estimated to be about 100 Wh/kg with a 1 h discharge rate in a thin-layer cell where LiC_6 is the negative electrode and a PANQ as the positive electrode [81]. The mean redox potential was determined to be about 2.6 V more positive than Li/Li^+ couple, which makes it a potential positive electrode for lithium metal or lithium-ion batteries. The PANI derivatives, PT-DMPD and PT-BZ, developed by Fors and coworkers, were investigated as cathode materials in Lithium-ion battery (LiB) technology [93]. Li-Coin cells devices with PT-DMPD or PT-BZ as the cathode delivered discharge capacities of 128 mAh/g and 97 mAh/g, which are 82% and 76% of the theoretical capacity for PT-DMPD and PT-BZ, respectively. However, the Coulombic efficiency (35% for PT-DMPD and 44% for PT-BZ) were relatively low. Additionally, the cycling performances were also poor with the discharge capacities for PT-DMPD and PT-BZ after 50 cycles being 82 mAh/g (64% retention) and 64 mAh/g (66% retention), respectively. To improve the performance of the polymers, Pt-DMPD was copolymerized with a second phenothiazine unit that was crosslinked to another one through a N-atom. The amount of cross-linked monomer was varied. Improved Coulombic efficiencies (150 mAh/g) at very positive operating voltages (2.8–4.3 V vs. Li+/Li) were obtained, which yielded high energy densities. Also, a greater retention capacity (82%) was observed at ultrafast discharge rates (120 C) for the crosslinked co-polymers.

4. Conclusion

To conclude, PANI derivatives are being widely developed in an attempted to address the limitations of PANI, which are insolubility and electrochemical instability. This short summary outlined both older but includes many new developments of PANI derivatives that are able to sufficiently overcome PANI's challenges and in some cases outperform PANI in several applications. Interestingly, with the developments of new synthetic approach by the Buchwald/Hartwig reaction to synthesize PANI derivatives, many new structures can be achieved that are not possible with the previous synthetic methods of oxidative and electrochemical polymerization. Consequently, it serves to continue developing new PANI derivatives by expanding the substrate scopes to continue to move the field forward in improving the current applications and merge into newer fields.

Acknowledgements

The authors are grateful for the financial support from the National Science Foundation for an award (CHE-1945503).

Conflict of interest

"The authors declare no conflict of interest."

Author details

Hari Giri[1], Timothy J. Dowell[1], Mohammed Almtiri[2] and Colleen N. Scott[1*]

1 Mississippi State University, Mississippi State, MS, USA

2 King Saud Bin Abdulaziz University for Health Sciences, Saudi Arabia

*Address all correspondence to: cscott@chemistry.msstate.edu

References

[1] MacDiarmid AG. "Synthetic metals": A novel role for organic polymers (Nobel lecture). Angewandte Chemie International Edition. 2001;**40**(14):2581-2590

[2] Basescu N, Liu Z-X, Moses D, Heeger AJ, Naarmann HD, Theophilou N. High electrical conductivity in doped polyacetylene. Nature. 1987;**327**:403-405

[3] Shirakawa H, Louis EJ, MacDiarmid AG, Chiang CK, Heeger AJ. Synthesis of electrically conducting organic polymers: Halogen derivatives of polyacetylene, (CH). Journal of the Chemical Society, Chemical Communications. 1977;**16**:578-580

[4] Swager TM. 50th anniversary perspective: Conducting/semiconducting conjugated polymers. A personal perspective on the past and the future. Macromolecules. 2017;**50**(13):4867-4886

[5] Guo X, Facchetti A. The journey of conducting polymers from discovery to application. Nature Materials. 2020;**19**(9):922-928

[6] Rasmussen SC. Conjugated and conducting organic polymers: The first 150 years. ChemPlusChem. 2020;**85**(7):1412-1429

[7] Tang H, Liang Y, Liu C, Hu Z, Deng Y, Guo H, et al. A solution-processed n-type conducting polymer with ultrahigh conductivity. Nature. 2022;**611**(7935):271-277

[8] Baker CO, Huang X, Nelson W, K aner RB. Polyaniline nanofibers: Broadening applications for conducting polymers. Chemical Society Reviews. 2017;**46**(5):1510-1525

[9] Sinha S, Bhadra S, Khastgir D. Effect of dopant type on the properties of polyaniline. Journal of Applied Polymer Science. 2009;**112**(5):3135-3140

[10] Yadav A, Kumar H. Polyaniline plastic nanocomposite as multi-functional nanomaterial. ChemistrySelect. 2022;**7**(29):e202201475

[11] Cao Y, Smith P, Heeger AJ. Counter-ion induced processibility of conducting polyaniline and of conducting polyblends of polyaniline in bulk polymers. Synthetic Metals. 1992;**48**(1):91-97

[12] Rasmussen SC. The early history of polyaniline: Discovery and origins. Substantia. 2017;**1**(2):99-109

[13] Majeed AH, Mohammed LA, Hammoodi OG, Sehgal S, Alheety MA, Saxena KK, et al. A review on polyaniline: Synthesis, properties, nanocomposites, and electrochemical applications. International Journal of Polymer Science. 2022;**2022**:9047554

[14] Das TK, Prusty S. Review on conducting polymers and their applications. Polymer-Plastics Technology and Engineering. 2012;**51**(14):1487-1500

[15] Taka T. Humidity dependency of electrical conductivity of doped polyaniline. Synthetic Metals. 1993;**57**(2):5014-5019

[16] Jin J, Wang Q, Haque MA. Length-scale effects on electrical and thermal transport in polyaniline thin films. Organic Electronics. 2010;**11**(1):29-35

[17] Campos M, Bulhões LOS, Lindino CA. Gas-sensitive characteristics of metal/semiconductor polymer

Schottky device. Sensors and Actuators A: Physical. 2000;**87**(1):67-71

[18] Beygisangchin M, Abdul Rashid S, Shafie S, Sadrolhosseini AR, Lim HN. Preparations, properties, and applications of polyaniline and polyaniline thin films—A review. Polymers. 2021;**13**(12):2003

[19] Mustafin AG, Latypova LR, Andriianova AN, Mullagaliev IN, Salikhov SM, Salikhov RB, et al. Polymerization of new aniline derivatives: Synthesis, characterization and application as sensors. RSC Advances. 2021;**11**(34):21006-21016

[20] Eftekhari A, Li L, Yang Y. Polyaniline supercapacitors. Journal of Power Sources. 2017;**347**:86-107

[21] Rangel-Olivares FR, Arce-Estrada EM, Cabrera-Sierra R. Synthesis and characterization of polyaniline-based polymer nanocomposites as anti-corrosion coatings. Coatings. 2021;**11**(6):653

[22] Zeng M, Zhu W, Luo J, Song N, Li Y, Chen Z, et al. Highly efficient nonfullerene organic solar cells with a self-doped water-soluble neutral polyaniline as hole transport layer. Solar RRL. 2021;**5**(3):2000625

[23] Gupta TK, Singh BP, Mathur RB, Dhakate SR. Multi-walled carbon nanotube–graphene–polyaniline multiphase nanocomposite with superior electromagnetic shielding effectiveness. Nanoscale. 2014;**6**(2):842-851

[24] Neoh KG, Pun MY, Kang ET, Tan KL. Polyaniline treated with organic acids: Doping characteristics and stability. Synthetic Metals. 1995;**73**(3):209-215

[25] MacDiarmid AG, Epstein AJ. Polyanilines: A novel class of conducting polymers. Faraday Discussions of the Chemical Society. 1989;**88**(0):317-332

[26] Liao G, Li Q, Xu Z. The chemical modification of polyaniline with enhanced properties: A review. Progress in Organic Coatings. 2019;**126**:35-43

[27] Malhotra B, Dhand C, Lakshminarayanan R, Dwivedi N, Mishra S, Solanki P, et al. Polyaniline-based biosensors. Nanobiosensors in Disease Diagnosis. 2015;**4**:25

[28] MacDiarmid AG, Epstein AJ. The concept of secondary doping as applied to polyaniline. Synthetic Metals. 1994;**65**(2):103-116

[29] Kuzmany H, Sariciftci NS, Neugebauer H, Neckel A. Evidence for two separate doping mechanisms in the polyaniline system. Physical Review Letters. 1988;**60**(3):212-215

[30] Wang L, Jing X, Wang F. On the iodine-doping of polyaniline and poly-ortho-methylaniline. Synthetic Metals. 1991;**41**(1):739-744

[31] Huang WS, Angelopoulos M, White JR, Park JM. Metallization of printed circuit boards using conducting polyaniline. Molecular Crystals and Liquid Crystals Incorporating Nonlinear Optics. 1990;**189**(1):227-235

[32] Ravichandran R, Sundarrajan S, Venugopal JR, Mukherjee S, Ramakrishna S. Applications of conducting polymers and their issues in biomedical engineering. Journal of the Royal Society Interface. 2010;7(suppl_5):S559-SS79

[33] Bhadra S, Chattopadhyay S, Singha NK, Khastgir D. Improvement of conductivity of electrochemically synthesized polyaniline. Journal

of Applied Polymer Science. 2008;**108**(1):57-64

[34] Bhadra S, Singha NK, Khastgir D. Polyaniline by new miniemulsion polymerization and the effect of reducing agent on conductivity. Synthetic Metals. 2006;**156**(16):1148-1154

[35] Bhadra S, Singha NK, Khastgir D. Effect of aromatic substitution in aniline on the properties of polyaniline. European Polymer Journal. 2008;**44**(6):1763-1770

[36] Bednarczyk K, Matysiak W, Tański T, Janeczek H, Schab-Balcerzak E, Libera M. Effect of polyaniline content and protonating dopants on electroconductive composites. Scientific Reports. 2021;**11**(1):7487

[37] Sapurina IY, Shishov MA. Oxidative polymerization of aniline: Molecular synthesis of polyaniline and the formation of supramolecular structures. In: Ailton De Souza G, editor. New Polymers for Special Applications. Rijeka: IntechOpen; 2012. p. 9

[38] Tang S-J, Wang A-T, Lin S-Y, Huang K-Y, Yang C-C, Yeh J-M, et al. Polymerization of aniline under various concentrations of APS and HCl. Polymer Journal. 2011;**43**(8):667-675

[39] Sapurina I, Stejskal J. The mechanism of the oxidative polymerization of aniline and the formation of supramolecular polyaniline structures. Polymer International. 2008;**57**(12):1295-1325

[40] Ćirić-Marjanović G. Recent advances in polyaniline research: Polymerization mechanisms, structural aspects, properties and applications. Synthetic Metals. 2013;**177**:1-47

[41] Yasuda A, Shimidzu T. Chemical and electrochemical analyses of polyaniline prepared with FeCl3. Synthetic Metals. 1993;**61**(3):239-245

[42] Mezhuev YO, Korshak YV. Theory of chain growth in chemical oxidative polymerization of aniline derivatives. Synthetic Metals. 2020;**267**:116445

[43] Gospodinova N, Terlemezyan L. Conducting polymers prepared by oxidative polymerization: Polyaniline. Progress in Polymer Science. 1998;**23**(8):1443-1484

[44] Wolfe JP, Wagaw S, Marcoux J-F, Buchwald SL. Rational development of practical catalysts for aromatic carbon–nitrogen bond formation. Accounts of Chemical Research. 1998;**31**(12):805-818

[45] Hartwig JF. Transition metal catalyzed synthesis of arylamines and aryl ethers from aryl halides and triflates: Scope and mechanism. Angewandte Chemie International Edition. 1998;**37**(15):2046-2067

[46] Yang BH, Buchwald SL. Palladium-catalyzed amination of aryl halides and sulfonates. Journal of Organometallic Chemistry. 1999;**576**(1):125-146

[47] Zhang X-X, Sadighi JP, Mackewitz TW, Buchwald SL. Efficient synthesis of well-defined, high molecular weight, and processible polyanilines under mild conditions via palladium-catalyzed amination. Journal of the American Chemical Society. 2000;**122**(31):7606-7607

[48] Ward RE, Meyer TY. O,p-Polyaniline: A new form of a classic conducting polymer. Macromolecules. 2003;**36**(12):4368-4373

[49] Spetseris N, Ward RE, Meyer TY. Linear and hyperbranched m-polyaniline: Synthesis of polymers for the study of magnetism in

organic systems. Macromolecules. 1998;**31**(9):3158-3161

[50] Ito A, Ota K-i, Tanaka K, Yamabe T, Yoshizawa K. n-alkyl group-substituted poly(m-aniline)s: Syntheses and magnetic properties. Macromolecules. 1995;**28**(16):5618-5625

[51] Goodson FE, Hartwig JF. Regiodefined poly(N-arylaniline)s and donor–acceptor copolymers via palladium-mediated amination chemistry. Macromolecules. 1998;**31**(5):1700-1703

[52] Casado N, Hernández G, Sardon H, Mecerreyes D. Current trends in redox polymers for energy and medicine. Progress in Polymer Science. 2016;**52**:107-135

[53] Geng Y, Jing X, Wang F. Solution properties of doped polyaniline. Journal of Macromolecular Science, Part B. 1997;**36**(1):125-135

[54] Rannou P, Nechtschein M, Travers JP, Berner D, Woher A, Djurado D. Ageing of PANI: Chemical, structural and transport consequences. Synthetic Metals. 1999;**101**(1):734-737

[55] Marmisollé WA, Inés Florit M, Posadas D. Electrochemically induced ageing of polyaniline. An electrochemical impedance spectroscopy study. Journal of Electroanalytical Chemistry. 2012;**673**:65-71

[56] Kobayashi T, Yoneyama H, Tamura H. Polyaniline film-coated electrodes as electrochromic display devices. Journal of Electroanalytical Chemistry and Interfacial Electrochemistry. 1984;**161**(2):419-423

[57] Liao G. The chemical modification of polyaniline with enhanced properties: A review. Progress in Organic Coatings. 2018;**126**:35-43

[58] Lindfors T, Ivaska A. pH sensitivity of polyaniline and its substituted derivatives. Journal of Electroanalytical Chemistry. 2002;**531**(1):43-52

[59] Park YW, Moon JS, Bak MK, Jin JI. Electrical properties of polyaniline and substituted polyaniline derivatives. Synthetic Metals. 1989;**29**(1):389-394

[60] Athawale AA, Kulkarni MV. Polyaniline and its substituted derivatives as sensor for aliphatic alcohols. Sensors and Actuators B: Chemical. 2000;**67**(1):173-177

[61] D'Aprano G, Leclerc M, Zotti G, Schiavon G. Synthesis and characterization of polyaniline derivatives: Poly(2-alkoxyanilines) and poly(2,5-dialkoxyanilines). Chemistry of Materials. 1995;7(1):33-42

[62] Ma L, Huang C-Q, Gan M-Y. Synthesis and anticorrosion properties of poly(2,3-dimethylaniline) doped with phosphoric acid. Journal of Applied Polymer Science. 2013;**127**(5):3699-3704

[63] Ma L, Wei S-J, Gan M-Y, Chen F-F, Tian N, Zeng J. Rapid emuision polymerization of poly(2,3-dimethylaniline) initiated by APS/Fe2+ composite oxidants and its performances. Acta Polymerica Sinica. 2013;**8**:1033-1038

[64] Patil S, Sainkar SR, Patil PP. Poly(o-anisidine) coatings on copper: Synthesis, characterization and evaluation of corrosion protection performance. Applied Surface Science. 2004;**225**(1):204-216

[65] Sathiyanarayanan S, Dhawan SK, Trivedi DC, Balakrishnan K. Soluble conducting poly ethoxy aniline as an inhibitor for iron in HCl. Corrosion Science. 1992;**33**(12):1831-1841

[66] Savitha P, Sathyanarayana DN. Synthesis and characterization of soluble

conducting poly(o−/m-toluidine-co-o-nitroaniline). Synthetic Metals. 2004;**145**(2):113-118

[67] Vicentini DS, Salvatierra RV, Zarbin AJG, Dutra LG, Sá MM. Synthesis and characterization of carboxyl-substituted polyanilines doped with halogenated acids: Combining conductivity with solubility. Journal of the Brazilian Chemical Society. 2014;**25**:1939-1947

[68] Yue J, Epstein AJ. Synthesis of self-doped conducting polyaniline. Journal of the American Chemical Society. 1990;**112**(7):2800-2801

[69] Milakin KA, Morávková Z, Taboubi O, Acharya U, Pop-Georgievski O, Bober P. Facile preparation of water-dispersible carboxylated polyaniline. Synthetic Metals. 2023;**293**:117249

[70] Karyakin AA, Strakhova AK, Yatsimirsky AK. Self-doped polyanilines electrochemically active in neutral and basic aqueous solutions: Electropolymerization of substituted anilines. Journal of Electroanalytical Chemistry. 1994;**371**(1):259-265

[71] Komkova MA, Valeev RG, Kolyagin YG, Andreev EA, Beltukov AN, Nikitina VN, et al. Solid-state survey of boronate-substituted polyaniline: On the mechanism of conductivity, electroactivity, and interactions with polyols. Materials Today Chemistry. 2022;**26**:101070

[72] Yue J, Epstein AJ. XPS study of self-doped conducting polyaniline and parent systems. Macromolecules. 1991;**24**(15):4441-4445

[73] Yue J, Wang ZH, Cromack KR, Epstein AJ, MacDiarmid AG. Effect of sulfonic acid group on polyaniline backbone. Journal of the American Chemical Society. 1991;**113**(7):2665-2671

[74] Saidu FK, Joseph A, Varghese EV, Thomas GV. Characterization and electrochemical studies on poly(1-naphthylamine)-graphene oxide nanocomposites prepared by in situ chemical oxidative polymerization. Journal of Solid State Electrochemistry. 2019;**23**(10):2897-2906

[75] Jadoun S, Verma A, Ashraf SM, Riaz U. A short review on the synthesis, characterization, and application studies of poly(1-naphthylamine): A seldom explored polyaniline derivative. Colloid and Polymer Science. 2017;**295**(9):1443-1453

[76] Faria RC, Bulhões LOS. Synthesis and electrochemical response of poly-(1-aminoanthracene) films. Electrochimica Acta. 1999;**44**(10):1597-1605

[77] Faria RC, Bulhões LOS. Hydrogen ion selective electrode based on poly(1-aminoanthracene) film. Analytica Chimica Acta. 1998;**377**(1):21-27

[78] Gou P, Kraut ND, Feigel IM, Bai H, Morgan GJ, Chen Y, et al. Carbon nanotube Chemiresistor for wireless pH sensing. Scientific Reports. 2014;**4**(1):4468

[79] Li X-G, Huang M-R, Hua Y-M, Zhu M-F, Chen Q. Facile synthesis of oxidative copolymers from aminoquinoline and anisidine. Polymer. 2004;**45**(14):4693-4704

[80] Naoi K, Suematsu S, Manago A. Electrochemistry of poly(1,5-diaminoanthraquinone) and its application in electrochemical capacitor materials. Journal of The Electrochemical Society. 2000;**147**(2):420

[81] Häringer D, Novák P, Haas O, Piro B, Pham MC. Poly(5-amino-1,4-naphthoquinone), a novel

lithium-inserting electroactive polymer with high specific charge. Journal of The Electrochemical Society. 1999;**146**(7):2393

[82] Pham MC, Piro B, Bazzaoui EA, Hedayatullah M, Lacroix J-C, Novák P, et al. Anodic oxidation of 5-amino-1,4-naphthoquinone (ANQ) and synthesis of a conducting polymer (PANQ). Synthetic Metals. 1998;**92**(3):197-205

[83] Zhao L, Wang W, Wang A, Yuan K, Chen S, Yang Y. A novel polyquinone cathode material for rechargeable lithium batteries. Journal of Power Sources. 2013;**233**:23-27

[84] Ismail KM, Khalifa ZM, Azzem MA, Badawy WA. Electrochemical preparation and characterization of poly(1-amino-9,10-anthraquinone) films. Electrochimica Acta. 2002;**47**(12):1867-1873

[85] Yano J, Ota Y, Kitani A. Electrochemical preparation of conductive poly(N-alkylaniline) s with long N-alkyl chains using appropriate dopant anions and organic solvents. Materials Letters. 2004;**58**(12):1934-1937

[86] Hwang G-W, Wu K-Y, Hua M-Y, Lee H-T, Chen S-A. Structures and properties of the soluble polyanilines, N-alkylated emeraldine bases. Synthetic Metals. 1998;**92**(1):39-46

[87] Nguyen MT, Kasai P, Miller JL, Diaz AF. Synthesis and properties of novel water-soluble conducting polyaniline copolymers. Macromolecules. 1994;**27**(13):3625-3631

[88] DeArmitt C, Armes SP, Winter J, Uribe FA, Gottesfeld S, Mombourquette C. A novel N-substituted polyaniline derivative. Polymer. 1993;**34**(1):158-162

[89] Hany P, Geniès EM, Santier C. Polyanilines with covalently bonded alkyl sulfonates as doping agent. Synthesis and properties. Synthetic Metals. 1989;**31**(3):369-378

[90] Stille JK, Mainen EL. Thermally stable ladder polyquinoxalines. Macromolecules. 1968;**1**(1):36-42

[91] Kim O-K. Electrical conductivity of heteroaromatic ladder polymers. 3. Phenothiazine and the structurally related ladder polymers. Journal of Polymer Science: Polymer Letters Edition. 1985;**23**(3):137-139

[92] Wu J, Rui X, Long G, Chen W, Yan Q, Zhang Q. Pushing up lithium storage through nanostructured Polyazaacene analogues as anode. Angewandte Chemie International Edition. 2015;**54**(25):7354-7358

[93] Peterson BM, Ren D, Shen L, Wu Y-CM, Ulgut B, Coates GW, et al. Phenothiazine-based polymer cathode materials with ultrahigh power densities for lithium ion batteries. ACS Applied Energy Materials. 2018;**1**(8):3560-3564

[94] Peterson BM, Shen L, Lopez GJ, Gannett CN, Ren D, Abruña HD, et al. Elucidation of the electrochemical behavior of phenothiazine-based polyaromatic amines. Tetrahedron. 2019;**75**(32):4244-4249

[95] Almtiri M, Dowell TJ, Chu I, Wipf DO, Scott CN. Phenoxazine-containing polyaniline derivatives with improved electrochemical stability and processability. ACS Applied Polymer Materials. 2021;**3**(6):2988-2997

[96] Almtiri M, Dowell TJ, Giri H, Wipf DO, Scott CN. Electrochemically stable carbazole-derived polyaniline for Pseudocapacitors. ACS Applied Polymer Materials. 2022;**4**(5):3088-3097

[97] Qi Y, Almtiri M, Giri H, Jha S, Ma G, Shaik AK, et al. Evaluation of the passivation effects of PEDOT: PSS on inverted perovskite solar cells. Advanced Energy Materials;**n/a**(n/a):2202713

[98] Orlov AV, Kiseleva SG, Karpacheva GP, Muratov DG. Peculiarities of oxidative polymerization of Diarylaminodichlorobenzoquinones. Polymers. 2021;**13**(21):3657

[99] Andriianova AN, Latypova LR, Vasilova LY, Kiseleva SV, Zorin VV, Abdrakhmanov IB, et al. Antibacterial properties of polyaniline derivatives. Journal of Applied Polymer Science. 2021;**138**(47):51397

[100] Robertson J, Gizdavic-Nikolaidis M, Nieuwoudt MK, Swift S. The antimicrobial action of polyaniline involves production of oxidative stress while functionalisation of polyaniline introduces additional mechanisms. PeerJ. 2018;**6**:e5135

[101] Quan X, Wang J, Souleyman T, Cai W, Zhao S, Wang Z. Antibacterial and antifouling performance of bisphenol-a/poly(ethylene glycol) binary epoxy coatings containing bromine-benzyl-disubstituted polyaniline. Progress in Organic Coatings. 2018;**124**:61-70

[102] Rai R, Roether JA, Boccaccini AR. Polyaniline based polymers in tissue engineering applications: A review. Progress in Biomedical Engineering. 2022;**4**(4):042004

[103] Wang L, Feng F, Ma Z. Novel electrochemical redox-active species: One-step synthesis of polyaniline derivative-Au/Pd and its application for multiplexed immunoassay. Scientific Reports. 2015;**5**(1):16855

[104] Borole DD, Kapadi UR, Mahulikar PP, Hundiwale DG. Glucose oxidase electrodes of polyaniline, poly(o-anisidine) and their co-polymer as a biosensor: A comparative study. Journal of Materials Science. 2007;**42**(13):4947-4953

[105] Lu Q, Qian J, Xu F, He G, Liu Y, Xia J. Synthesis of 2D fluorescent polyaniline derivatives as multifunctional fluorescent chemosensor. Journal of Polymer Science. 2023;**61**(3):211-222

[106] Qian J, Zhang Y, Liu X, Xia J. Carbazole and fluorene polyaniline derivatives: Synthesis, properties and application as multiple stimuli-responsive fluorescent chemosensor. Talanta. 2019;**204**:592-601

[107] Prakash R, Santhanam KSV. Electrochromic window based on polyaniline. Journal of Solid State Electrochemistry. 1998;**2**(2):123-125

[108] Saharan R, Kaur A, Dhawan SK. Synthesis and characterization of poly(o-methoxy aniline) and its copolymer for electrochromic device energy applications. Indian Journal of Pure & Applied Physics. 2015;**53**:316-319

[109] Manisankar P, Vedhi C, Selvanathan G, Somasundaram RM. Electrochemical and electrochromic behavior of novel poly(aniline-co-4,4′-diaminodiphenyl sulfone). Chemistry of Materials. 2005;**17**(7):1722-1727

[110] Marques AS, Szostak R, Marchezi PE, Nogueira AF. Perovskite solar cells based on polyaniline derivatives as hole transport materials. Journal of Physics: Energy. 2019;**1**(1):015004

[111] Brusic V, Angelopoulos M, Graham T. Use of polyaniline and its derivatives in corrosion protection of copper and silver. Journal of The Electrochemical Society. 1997;**144**(2):436

[112] Zhou Z, Xiao X, Wang W, Wei S, Wang Y. Enhanced hydrophobicity and barrier property of anticorrosive coatings with silicified polyaniline filler. Colloids and Surfaces A: Physicochemical and Engineering Aspects. 2022;**645**:128848

[113] Xing C, Song X, Zhang Z, Jiang X, Yu L. Anticorrosion coatings from poly (aniline-co-2-Ethylaniline) micro/nanostructures. Journal of Ocean University of China. 2019;**18**(6):1371-1381

[114] Jafari Y, Shabani-Nooshabadi M, Ghoreishi SM. Poly(2-chloroaniline) Electropolymerization coatings on Aluminum alloy 3105 and evaluating their corrosion protection performance. Transactions of the Indian Institute of Metals. 2014;**67**(4):511-520

[115] Cai W, Wang J, Quan X, Zhao S, Wang Z. Antifouling and anticorrosion properties of one-pot synthesized dedoped bromo-substituted polyaniline and its composite coatings. Surface and Coatings Technology. 2018;**334**:7-18

[116] Ahmad S, Ashraf SM, Riaz U, Zafar S. Development of novel waterborne poly(1-naphthylamine)/poly(vinylalcohol)–resorcinol formaldehyde-cured corrosion resistant composite coatings. Progress in Organic Coatings. 2008;**62**(1):32-39

[117] Riaz U, Khan S, Islam MN, Ahmad S, Ashraf SM. Evaluation of antibacterial activity of nanostructured poly(1-naphthylamine) and its composites. Journal of Biomaterials Science. Polymer Edition. 2008;**19**(11):1535-1546

Chapter 3

Recent Developments in the Use of Polyaniline-Based Materials for Electric and Magnetic Field Responsive Smart Fluids

Ozlem Erol

Abstract

Smart fluids are stimuli-responsive materials whose rheological properties can be changed drastically by applying either an external electric or magnetic field strength. Smart fluids are dispersions comprised of dispersed particles in a carrier liquid that transform from liquid-like state to solid-like state within milliseconds reversibly with an application of external field due to the structural chain formation of the dispersed particles. Owing to this outstanding controllable transformation capability, smart fluids are utilized in various potential applications where an electro/magneto-mechanical interface is required, such as dampers, clutches, shock absorbers, robotics, haptic devices, microfluidics, etc. Various kinds of materials have been proposed and used by researchers for applications that require the electrorheological (ER) and magnetorheological (MR) effects. Polyaniline (PAn) is considered a remarkable material as a dispersed phase of ER fluids due to its easy synthesis, low cost, adjustable conductivity through doping/de-doping processes, and excellent environmental stability. PAn is an attractive material in MR fluids as well due to its contribution to the improvement of dispersion stability and protection against corrosion and oxidation of the soft-magnetic particles. In this chapter, the recent advances in the usage of various kinds of PAn-based materials as electric and magnetic field responsive materials and their ER/MR behaviors are summarized.

Keywords: polyaniline, polyaniline-based-nanocomposites, polyaniline-based-hybrid materials, electrorheological fluids, magnetorheological fluids

1. Introduction

Electrorheological (ER) and magnetorheological (MR) fluids are stimuli-responsive smart dispersions with rheological properties (yield stress, viscosity, shear modulus, etc.) that are reversible and controllable in a continuous manner with the usage of an externally applied electric field (E) or magnetic field (H), respectively. The ER fluids are commonly composed of polarizable/semiconducting micron/nano-sized particles as a dispersing phase, in an insulating dispersing medium with low volatility and high chemical/thermal stability, whereas their magnetic counterpart, the MR fluids generally consist of highly magnetizable micron-sized dispersed particles in a nonmagnetizable carrier fluid such as mineral oil, silicone oil, polyesters, polyethers, synthetic hydrocarbons, or water. Additives can also be added to enhance the ER/MR effect and/or to increase dispersion stability of the whole stimuli-responsive fluids. These smart fluids transform reversibly and rapidly from a liquid-like state to a solid-like state within a millisecond with the aid of an E or a H. Smart fluids, therefore, can be used as electrical and mechanical interfaces in various applications that require active control of vibrations or the transmission of torque, including the fast-acting valves, clutches, brakes, shock absorbers, accurate polishing, robotics, and tactile displays [1, 2].

For ER fluids, when an E is applied, the randomly dispersed particles are polarized forming a dipole moment due to the difference between the dielectric constant of the dispersant and dispersed particles. These particles attract each other along the field between the parallel electrodes to construct chain and/or columnar-like structures along the field direction. Similarly, when MR fluids are exposed to a H, dispersed particles are magnetized and oriented along the direction of the H and generate anisotropic aggregates. As a result, for both smart fluids, the field-induced structuration in the fluid leads to resist the flow of the carrier fluid, which results in an enhanced apparent viscosity and viscoelasticity of the fluid. This phenomenon is known as the ER or MR effect. The schematic representation of the microstructure of ER or MR fluids under on/off state of the E or H is shown in the **Figure 1**.

A deeper understanding of the field-induced structuring mechanism may lead us to design and prepare high-performance field responsive fluids. The mechanism of the electric field-induced rheological changes has been studied intensively since the invention of the ER fluids [3]. Several mechanisms or models were proposed previously to explain the ER effect, including fibrillation, electrical double layer, water/surfactant bridge, polarization, conduction, and dielectric loss model [4]. On the other hand, particle magnetization model is the most acceptable mechanism to account for the magnetic field-induced structuring in MR fluids. According to the particle magnetization model, the MR effect is ascribed to the magnetic permeability mismatch between the dispersed particles and continuous phases [2]. There are several critical key factors that influence the behavior of the ER fluids. Thus, it is difficult to develop desired high-performance ER fluids in consideration of all variables such as dispersed particle size, shape and overall morphology, particle conductivity and dielectric properties, particle surface properties, particle volume fraction of the ER fluid, the E,

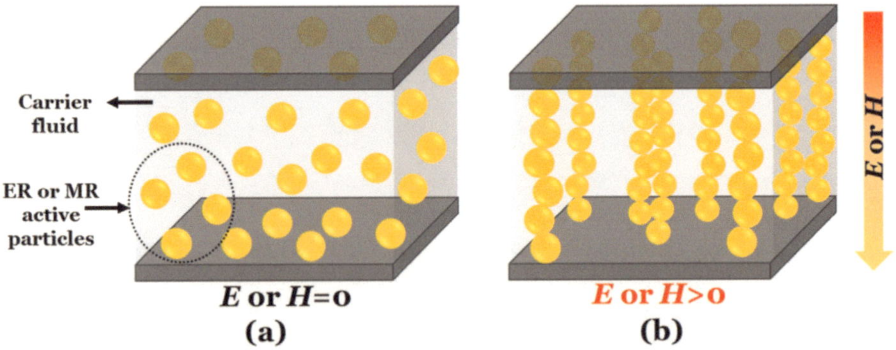

Figure 1.
Schematic representation of the microstructure of ER or MR fluids under off-state (a) and on-state (b) of E or H.

temperature, properties of dispersing medium, and dispersion stability. A proper electrical conductivity range is desired to be 10^{-6}–10^{-12} S/cm [5]. For MR fluids, the H, carrier fluid, additives, temperature, dispersion stability, durability, magnetic permeability, particle surface roughness, volume fraction, density, and size/shape of the dispersed particles are the major parameters on MR effect [6].

As a dispersed phase, a wide range of ER or MR active materials are developed to eliminate the impediments that limit the use of stimuli-responsive fluids in the literature. Various polarizable particles ranging from ceramics to polymers have been applied as the dispersed phase of ER fluids, including SiO_2, TiO_2, corn starch, aluminosilicate, carbonaceous particles, conducting polymers, etc. [1]. In addition, recent efforts on various hybrid materials with different hierarchical morphologies have been made to improve the dispersion stability and polarizability. As a dispersed phase, Fe_3O_4, Fe_2O_3, iron particles, carbonyl iron (CI) particles, iron alloy particles (nickel-iron, cobalt-iron, magnetized stainless steel alloys), and ferrite nanoparticles were used to fabricate MR fluids [7]. MR active particles are desired to have large saturation magnetization and small coercivity value over a wide temperature range, high dispersion, and chemical stability [2]. Although these MR active materials have reasonable mechanical properties, they generally have lack of dispersion stability against settling and poor anticorrosion properties. Thus, surface modification, coating of magnetic particles, or the addition of different additives are usually required. The optimum particle size of dispersed phase is usually in the range of 0.1–10 μm.

Among conductive polymers, PAn has been widely used in the research on ER materials so far due to its possessing excellent chemical and environmental stability and tunable conductivity. Additionally, PAn has been utilized for enhancing the dispersion stability of MR fluids and obtaining dual ER/MR response. In this chapter, the recently developed PAn-based materials as electric and magnetic field responsive materials and their ER/MR behaviors are reviewed.

2. Polyaniline-based ER fluids

As an anhydrous ER active material, PAn and PAn-based materials have been extensively investigated in previous studies due to their adjustable electrical conductivity, easy synthesis, low cost, good environmental, thermal, and chemical stability, and nonabrasive for the device. Various efforts have been made to increase the ER efficiency of PAn, including synthetization of PAn with different morphologies, preparing its derivatives, and combining PAn with various kinds of materials such as inorganic particles or polymers by preparing composites, nanocomposites, core/shell structures, etc. In this section, constituents of PAn-based ER fluids have been reported so far and their general ER characteristics are summarized. The main parameters of PAn-based ER fluids can be found in a brief table in a recent review reported by Kuznetsov et al. [8].

2.1 Development and diversification of polyaniline-based ER fluids

2.1.1 Polyaniline, its derivatives, and copolymers

Among conjugated conductive polymers, PAn has drawn great attention due to its easy and cost-effective production. PAn is synthesized in different morphologies, including fibrillar, tubular, belt-like, porous, hollow structures, etc. PAn exists in different forms in terms of its oxidation state as fully oxidized, half-oxidized, and fully reduced states, namely, pernigraniline, emeraldine base, and leucomeraldine, respectively. For ER application, the level of conductivity of the fabricated PAn is important and depends on the degree of doping, synthesis conditions such as temperature, monomer: oxidatant ratio, presence of surfactant, and other components. If the conductivity of the PAn is not within the proper semiconductive level, dedoping process is generally applied to PAn before preparing ER fluids. The doping process by organic/inorganic acids also impacts the control of wettability of PAn, which plays a role in ER efficiency by influencing the compatibility between solid phase and liquid phase [9, 10].

The doping degree of PAn can be controlled by adjusting the pH of the aqueous solution containing the PAn particles to 9.0 to make it suitable for ER fluids. Additionally, Xie et al. reported the effect of different dedoping methods on ER properties of PAn, namely equilibrium and nonequilibrium. In the reported equilibrium method, the PAn particles were immersed in a diluted aqueous ammonia solution with a certain pH value and stirred for a defined time. On the other hand, in the reported nonequilibrium method, the PAn particles were immersed in a concentrated aqueous ammonia solution for a certain time. Although similar conductivity values were reached for both methods, the greater yield stress value was observed in case of the PAn prepared *via* nonequilibrium method. The reason for this observation was attributed to the possible formation of more insulating shell coating on PAn particles *via* nonequilibrium method, which led to more electron movement within the particles rather than electron hopping between them [11]. However, when the PAn is even dedoped, the current density may still be higher than that of desired for ER fluid and cause consumption of power and electrical discharge. Thus, PAn derivatives and copolymers have taken attention owing to eliminate the limitations of PAn.

Substituted derivatives of PAn are obtained by the introducing of substituted groups on the benzene ring and/or amino N of PAn. The conjugation length and electrochemical behaviors are both affected by the substituent nature and its position on the ring [12]. The PAn derivatives usually show reduced intrinsic conductivities by several orders of magnitude compared to PAn due to steric hindrance of the substituent groups or the disruption of the conjugated backbone or difference in the interchain interactions [13]. Thus, further dedoping process, which is generally required for PAn-based ER fluids, may not be required for PAn derivatives, and the efficiency of the processing of ER fluid can be increased. The most extensively investigated introduced substituents are methyl, ethyl, methoxy, ethoxy, and phenyl groups. There are vast amount of different homopolymers and copolymers have been prepared from substituted anilines for ER application in the literature, including poly (2-methoxyaniline) or poly(*o*-anisidine) [14], poly(2-ethoxyaniline), poly(N-methylaniline), poly(N-ethylaniline), poly(2-methylaniline) or poly(*o*-toluidine) [15, 16], poly(2-ethylaniline) [16, 17], poly(diphenylamine) (N-aryl-substituted PAn) [18], substituted PAn with long alkyl pendants (poly(2-dodecyloxyaniline) [19], poly (aniline-*co*-diphenylamine) [20], poly(aniline-*co*-*o*-ethoxyaniline) [21], poly(aniline-*co*-1,4-phenylenediamine) [22], PAn copolymer containing N-substituted benzene sulfonic acid group [23], and aniline/pyrrole copolymer [24]. In addition, oligomeric aniline particles were demonstrated to exhibit greater ER activity than PAn [25].

An irregular particulate morphology is generally obtained from chemical oxidative polymerization of aniline and used as ER active material. Additionally, well-defined hollow [26], nanofibrous [27, 28], nanotubular [29], nanorod covered rectangular tubular [30], urchin-like [31], and clip-like [32] shaped PAn have been developed using a variety of synthesis methods, such as template synthesis, rapid mixing polymerization or interface polymerization, chemical oxidation dilute polymerization in an aqueous acidic solution with an anionic surfactant, the *in-situ* chemical oxidative polymerization assisted with a surfactant mixed system. In a study, the effect of morphology and size, that is nanofiber, nanoparticle, and microparticle PAn prepared by modified oxidative polymerization method in the citric acid aqueous solution, on the sedimentation and ER properties of PAn has been reported the highest dispersion stability and the strongest ER effect were observed for nanofiber morphology compared to that of the nanoparticle and microparticle ones due to its possession of high aspect ratio [33]. In another study, porous PAn was prepared *via* chemical oxidation polymerization at various synthesis temperatures between 37°C to 95°C and reduced pore size, increased porosity, and specific surface area, and stronger ER response with lower leakage current density were reported with increasing synthesis temperature. It has been stated that the increased specific surface area and porosity lead to an increase in the interaction area between the carrier fluid and the PAn particles in the ER fluid, resulting in enhanced interfacial polarization and ER performance [34]. On the other hand, the highest ER activity was reported for the dispersion of particles obtained at the lowest synthesis temperature due to possessing higher molar mass and longer main chains of PAn, at the same time, having greater the dielectric permittivity and dielectric loss coefficient has been observed in a different study where the effect of lower polymerization temperatures in the range from −10°C to 5°C was studied [35].

2.1.2 Polyaniline/inorganic hybrid materials

PAn and PAn derivatives have been hybridized with inorganic materials in order to achieve the synergistic contribution of constituent materials, which cannot be provided from one another, to improve ER characteristics and dispersion stability of the fluid and adjust the conductivity in a desired range for ER applications [36]. Several methods have been applied to fabricate PAn/inorganic hybrid materials including *in-situ* chemical oxidative polymerization, emulsion polymerization, microemulsion polymerization, and Pickering emulsion polymerization of PAn and PAn derivatives with the presence of inorganic material. As an inorganic part of the PAn/inorganic hybrid material, titanate [37, 38], titania [39, 40], TiO_2 [41–43], kaolinite [44], silica [14, 45–47], halloysite [48], sepiolite [49, 50], Fe_2O_3 [51, 52], Fe_3O_4 [53, 54], palygorskite [55], $BaTiO_3$ [56–58], bentonite [59], montmorillonite [60–62], organoclay [63, 64], organo-montmorillonite [65], laponite [66], mesoporous SiO_2 [67–70], mesoporous TiO_2 [71], anisotropic TiO_2 [72], attapulgite [73], nanoporous zeolite [74], metal-organic framework (MOF) [75], MoS_2 [76, 77], WS_2 [78], K-feldspar [79], red mud [80], aluminum hydroxide and talc [81], and zinc ferrite [82] have been used. Among these inorganic compounds, since silica is a well-defined uniform structured compound that can be fabricated on large scale, it has been widely studied as a typical ER active material. In addition, ER fluid containing mesoporous form of silica has demonstrated improved dispersion stability, dielectric property, and ER activity due to its additional advantages, such as low density and high surface area compared to bare silica [69]. However, since silica has low conductivity, its ER response is limited. Researchers have enhanced ER activity of silica by controlling its size and morphology, and *via* coating the silica with conducting polymers and carbonaceous materials using different methods. For example, Noh et al. fabricated vapor deposition polymerized PAn-coated mesoporous silica particles with different aspect ratios to investigate the particle geometry effect on ER activity and demonstrated that the ER performance of the dedoped PAn-coated mesoporous silica was significantly enhanced due to the increase of the aspect ratio of the silica core [70].

Among the natural clay minerals, inherent 1D nanostructured sepiolite, palygorskite (attapulgite), and halloysite have high aspect ratio and surface charge, large specific surface area, good mechanical strength, outstanding chemical, thermal and dimensional stability, low cost, abundant natural resources, and easy processing and have been hybridized with PAn to utilize their 1D shape effect. In another study, porous zeolite allowed the PAn to fill into microstructural pores resulting in complex nanostructures, leading PAn composite to have better dielectric properties and ER performance due to stronger polarization and fibrillar formation under E [74]. A thin conducting layer of PAn is also one of the best ways of reducing the cost of conducting polymers in these inorganic/PAn nanostructured particles.

2.1.3 Polyaniline/carbon-based nanoparticle composites

Carbon-based nanostructures such as carbon nanoparticles, multi-walled carbon nanotubes (MWCNT), graphene, and graphene oxide (GO) have been utilized for the preparation of ER materials due to their unique and versatile electrical and morphological properties. The carbon-based materials, which stand out with their high conductivity, have been combined with polymers or inorganic compounds in order to reach the appropriate conductivity range for ER studies. These studies were in the form of coating carbon-based material on polymer or inorganic material, or, on the contrary, coating carbon-based material with polymer or inorganic material. Among the carbon-based materials, GO sheets, considered as an oxidation state of graphene, have the most remarkable result since these materials possess relatively reduced electrical conductivity even without any posttreatment and good dispersion stability in carrier fluid due to the large amount of hydroxyl, carboxylic, and epoxide functional groups on its basal planes and edges [83, 84]. PAn has been combined with various carbon-based materials for ER studies, including MWCNT [85], GO [86–88], graphene [89, 90], and carbon particles [91], in addition, 2D carbonaceous particles were fabricated from annealing of PAn-coated GO sheets in vacuum [92] and core–shell particles of carbonized PAn base was coated with PAn base [93].

In a study, nitrogen-enriched carbonaceous nanotubes with large aspect ratio and tunable conductivity were prepared by heat treatment of PAn nanotubes at elevated temperatures and used as ER active material. Thus, with this approach, to achieve the desired high polarizability and dispersion stability for ER fluids, the high aspect ratio of 1D morphology has been maintained with appropriate electrical conductivity, which is difficult to achieve in highly conductive carbon-based nanostructures. Furthermore, nitrogen-enriched carbonaceous nanotubes displayed better dispersion stability and higher ER performance compared to their granular analogue [29]. In another study, to display the morphology and conductivity effect on ER performance of composite nanoplates, *in-situ* fabricated GO-supported PAn emeraldine salt nanoplates were treated either with ammonia or hydrazine to obtain nonconducting-GO-supported PAn base nanoplates (rGO/PAn) or conducting reduced-GO-supported PAn base nanoplates (GO/PAn), respectively. The dielectric and ER properties of both composite nanoplates, with the same morphology and shell property but with different core conductivities, were compared with granular-shaped PAn. In case of using conducting rGO as the core instead of nonconducting one enhanced the intensity and the rate of interfacial polarization of composite nanoplates and thus resulted in stronger ER response. Furthermore, both anisotropic nanoplate composites displayed much greater ER performance than that of PAn base [90]. This reported study clearly indicated that the anisotropic GO structure had an influential role in improved polarization and ER effect, whereas the conductive core gave further favorable contribution to ER effect. Apart from this study, PAn-coated GO sheet prepared with a similar approach mentioned above was subjected to heat treatment at 550°C under inert atmosphere to fabricate two-dimensional (2D) structures composed of graphene-supported amorphous carbon. The electric field-induced yield stress values of 2D-graphene-supported carbonaceous sheets were reported to be about three times greater than that of dispersion of pure carbonaceous particles obtained from annealing of PAn due to the increased polarization with the aid of the presence of the graphene core. Additionally, the interparticle friction and viscous drag force have been enhanced owing to the high-aspect-ratio of 2D-plate-like morphology [92]. Instead of amorphous carbon coating, PAn base coated on graphitic carbon prepared from carbonized PAn base particles at 650°C under inert atmosphere was used as a dispersed phase of ER fluid. This composite structure was reported to lower the interactions between the carrier liquid and the dispersed particles and enhance the ER efficiency compared with the dispersion of the bare PAn base and carbonized PAn base in silicone oil [93].

2.1.4 Polyaniline/polymer composites

For ER application, the conductivity of PAn has to be adjusted to desired level to prevent electrical short upon application of the *E*. There are various alternative strategies to maintain the electrical characteristics of PAn at a safe and proper level for ER purposes and enhance the ER performance, such as dedoping process, co-polymerization, encapsulating of PAn with polymers or inorganic materials with low conductivity, fabricating of core/shell hybrid structures, nanocomposites, composites, and blends. The conductivity of the PAn incorporated materials can be controlled in a wide range by altering the amount of PAn. The increasing PAn content increases the density and mobility of the charge carriers until the optimum saturation level is reached depending on the conducting network formation in the dielectric polymer matrix [94]. Many researchers have devoted themselves to combining PAn particles with an insulating polymer for ER studies. Especially, core/shell structured uniform polymer microspheres coated PAn with controlled thickness have been extensively studied to combine the morphological effect of core nonconducting polymer with the electrical properties of conducting PAn layer. As a general approach, monodispersed micron-sized PAn or PAn derivative composites are synthesized *via* oxidative polymerization of aniline or aniline derivative monomer in the presence of micron-sized polymer spheres such as poly(styrene) (PS) [95, 96], poly(methyl methacrylate) (PMMA) [97, 98], poly(glycidyl methacrylate) (PGMA)/PMMA [99], or poly(styrene-*co*-glycidyl methacrylate-co-divinylbenzene) [100]. The ER effect of the PAn-coated polymer core/shell composite particles containing dispersions depends on the PAn loading, the dispersed particle conductivity, and concentration [100]. In a study, among the homopolymer PAn, PMMA(core)/PAn(shell), and PAn(core)/PMMA (shell) particles, the ER effect of the PAn(core)/PMMA(shell) particles was demonstrated to be stronger than that of PMMA(core)/PAn(shell) particles whereas similar with that of homopolymer PAn. As a result, an insulating shell on conductive PAn core was found to be effective to obtain improved ER performance without dedoping process [97]. Additionally, the particle size of the insulating polymer core with similar PAn shell thickness and electrical properties have been reported to influence the ER performance. The larger particle size in the range from 1 to 10 μm resulted in the greater ER effect [101]. In the literature, besides the isotropic spherical PAn/insulating polymer core/shell structures, the fabrication of anisotropic PAn-coated snowman-like core/shell particles was reported *via* seed emulsion polymerization method [98]. To enhance the mechanical strength of the core/shell structure, the crosslinking agent like ethylene glycol dimethacrylate has also been introduced to the PMMA core before being coated by the conducting PAn derivative overlayer via grafting polymerization [102]. Another approach to produce semiconducting core/shell-type nonconducting polymer coated on PAn particles is *via* Pickering emulsion-type polymerization, using polymer particles as a solid polymeric surfactant, such as poly(divinylbenzene-*alt*-maleic anhydride) [103].

2.1.5 Polyaniline/poly(ionic liquid) composites

Organic polymer-based materials have been most frequently studied as ER materials compared to the inorganic ones due to their low density, soft texture, and relatively high ER activity. Among various organic polymer-based polymers, polyelectrolytes are more extensively applied in ER fluids due to their low cost, facile fabrication, and high ER response in the presence of promoter, such as a small amount of water or moisture. The small amount of water required for ER activation leads the traditional polyelectrolytes to face the electrical and thermal problems. To overcome these drawbacks, a novel anhydrous polyelectrolyte-based ER system based on hydrophobic poly(ionic liquid)s (PILs) has been developed recently [104].

PILs are a type of functional polyelectrolyte obtained by polymerization of ionic liquids, a typical organic molten salt bearing hydrophobic counterions at room temperature. PILs can exhibit strong ER activity and high dielectric polarizability without affinity to water due to comprising of high-density cation/anion counterions endowing polarized easily upon application of the *E*. Additionally, crosslinking can be applied to improve the mechanical and thermal properties of PILs due to possessing low glass transition temperature. The ER activity of PILs with linear backbone depends on the type and the nature of the hydrophobic counterions [105]. If the PILs are self-crosslinked, the ER effect rises with the length of the alkyl spacer due to enhanced ion mobility and induced interfacial polarization whereas the operating temperature range is narrowed resulting in current leakage at higher temperature [106]. Hydrophobic PILs can be prepared by direct polymerization of hydrophobic ionic liquid monomers or subsequent post-ion-exchange treatment after polymerization of hydrophilic ionic liquid monomers. Although, the post ion-exchange treatment is a facile procedure with higher yield, hydrophobic PILs prepared by post ion-exchange procedure are easily surface charged. In order to eliminate the surface charging and thus enhance the ER effect, the preparation of the composite particles composed of conducting PAn core encapsulated by PIL, poly(vinylbenzyl)trimethylammonium hexafluorophosphate (P[VBTMA]PF_6) was reported *via* post-ion-exchange procedure. The P[VBTMA]PF_6-capsulated PAn particles were demonstrated to have an enhanced ER effect due to partially suppressing the positively charged state of P[VBTMA]PF_6 particles by wrapping PAn into P[VBTMA]PF_6 when compared with that of pure PIL, pure PAn, and their simple mixture [107].

At higher *E* values, the irreversible leakage of mobile ions from particles into carrier liquid can be encountered for PIL-based ER fluids. Another approach to block ion leakage from PILs can be semiconducting PAn coating on the surface of ionically conductive PIL core since PAn mainly allows the transport of the electron or the hole.

In a study, Zheng et al. demonstrated that core/shell structured poly[2-(methacryloyloxy)ethyl] trimethylammonium bis(trifluoromethanesulfonyl) imide (P[MTMA][TFSI])/PAn fabricated *via* low-temperature interfacial polymerization of PAn on the P[MTMA][TFSI] microspheres limited the irreversible ion leakage of P[MTMA][TFSI] microspheres and improved the particle polarizability and ER effect [108].

The influence of oxidation state of PAn on the ER response of PIL/PAn composite composed of P[VBTMA][PF_6] as a matrix and different forms of semiconducting PAn as a filler, prepared by ion-exchange and subsequently treated by ammonia or hydrazine to obtain different forms of PAn filler, such as emeraldine salt, emeraldine base, and leucoemeraldine was studied by Zheng et al. Thus, the researchers could adjust the difference in polarization rate between filler and matrix and reported that when the closer their polarization rates were, the more stable the flow curve was in a broad shear rate region, and enhanced ER response was achieved only for the ammonia-treated P[VBTMA][PF_6]/PAn particles at room temperature. On the other hand, the flow curves of the hydrazine-treated P[VBTMA][PF_6]/PAn displayed more stable flow curve with increasing temperature [109]. In another study, ionic liquid crystal PAn prepared *via* microwave-assisted reaction using leucoemeraldine PAn as polymer skeleton, naphthalene disulfonic acids as ionic crosslinkers, and diisonicotinates as mesogenic groups demonstrated integrated behaviors of PIL, liquid crystal, and PAn-base and strong ER dependency on operating temperature [110]. Emeraldine base form of PAn was doped by the synthesized series of main-chain liquid-crystalline polymers (LCPs) with pendant sulfonic acid groups using biphenyl-4,40-diol, 6,7-dihydroxynaphthalene-2-sulfonic acid, and bis(4-(chlorocarbonyl)phenyl) decanedioate in a one-step esterification reaction. The resulting sulfonic acid-containing PAn–LCP ionomer dispersions showed better ER effect than PAn dispersions [111].

2.2 ER characteristics of polyaniline-based materials

Within the realm of ER fluids, the ER characteristics have been extensively examined by rheological measurements including steady shear, on/off switch test at constant shear rate, and oscillatory test under E and dielectric analysis in terms of understanding the behavior of ER fluid and response to external electrical stimuli. ER properties investigated are mainly yield stress, storage, and loss modulus under different E values. Without E, the shear stress of ER fluid usually increases linearly with increasing shear rate displaying Newtonian fluid behavior (**Figure 2a**). In the presence of the E, the shear stress remains constant over a low shear rate range and increases with increasing the shear rate in the high shear rate range according to the Bingham model (**Figure 2a**). The yield stress (τ_y) is defined as the minimum shear stress required to completely disrupt the field-induced solid-like structure under continuous shear. However, in many cases, Bingham model is not sufficient for describing the flow curves of ER fluids over a large range of the shear rate. Therefore, other models such as De Kee-Turcotte model, the Hershel-Bulkley model, Cho-Choi-Jhon (CCJ) model (**Figure 2a**), and Seo-Seo model have been applied [112]. The τ_y can be determined by above-mentioned models using experimental data. The correlation of the τ_y and applied electric field follows power law, $\tau_y \propto E^{\alpha}$. The value of index, α, is 1.5 or 2.0 corresponding to the conduction model and ideal polarization model, respectively [5]. On the other hand, in some cases, this relationship is not fully

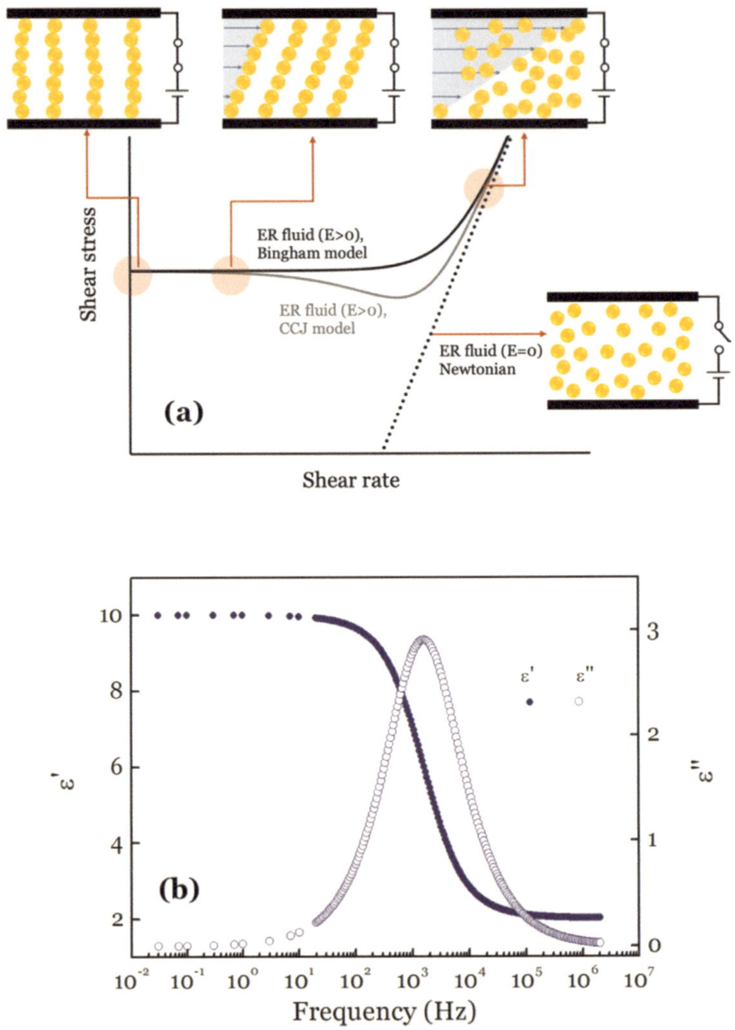

Figure 2.
Representative shear stress versus shear rate curves of an ER fluid that fits the Newtonian model under off-field, the Bingham or CCJ model under on-field, and corresponding schematic microstructure under shear flow (a) and representative dielectric spectrum (b) of an ER fluid.

consistent with the corresponding models due to the effect of dispersed particle size, morphology, surface properties and concentration, and dielectric properties of ER fluid on the τ_y.

The viscoelastic properties and phase transition of ER fluids can be examined by dynamic oscillation test in which a sinusoidal strain or stress is applied, and the ER fluid is sheared back and forth at a given strain amplitude and frequency [5]. In order to do so, a linear viscoelastic region (LVE), where stress and strain are proportional, is initially determined by stress or strain amplitude sweep test at a fixed frequency value. Within the low-strain region, elastic (G′) and viscous (G″) modulus are independent of the applied strain or stress and show a constant plateau. When the applied strain or stress is inadequate and leads to structural breakdown of the field-induced structures, the important microstructural properties are being measured. Out of the LVE, nonlinearities arise and measurements can no longer be easily correlated with microstructural properties. Within the predetermined LVE for the given ER fluid, the angular frequency sweep test is performed to analyze the time-dependent behavior of the ER fluid in the nondestructive deformation range and to determine the behavior and inner structure, and the long-term stability of the ER fluid [113]. In the absence of the *E*, G′<G″ is observed at low frequencies indicating liquid-like behavior predominates, then the crossover point of G′ and G″ is appeared, which is related to the relaxation time, and at high-frequency region G′>G″ is displayed indicating solid-like behavior predominates for viscoelastic liquids. Upon application of electric field, G′>G″ is displaced and G′ and G″ are observed as parallel throughout the entire frequency range indicating stable gel-like structure with solid-like behavior. With increasing the *E*, the modulus values increase indicating raised structural strength.

Furthermore, the microstructural changes in the ER fluid can be directly determined by optical microscopy under *E*. The particles are dispersed in the carrier fluid randomly whereas they rapidly align parallel to the direction of the *E* and form a chain-like structure spinning towards the electrodes. Depending on the concentration of the ER fluid and the magnitude of *E*, columnar or network structure can also be observed.

Dielectric spectra of ER fluids, the frequency-dependent dielectric constant (ε′) and dielectric loss factor (ε″) curves (**Figure 2b**), provide important information about the polarization mechanism, polarizability, and polarization relaxation time (λ) of ER fluids on analyzing of electrical polarization properties and interpreting the flow behavior under the *E* [13]. Response time of the dispersed particles to the *E* and formation of stable fibrillar structures are related to the relaxation frequency of the ER fluid. A high relaxation frequency results in a short relaxation time and a rapid response time to an *E*. Generally, a higher polarizability corresponds to strong interactions between particles and ER performance. For achieving stable flow behavior under an *E*, the response rate of the dispersed particles to *E* was presented in the range of 10^{-5}–10^{-2} s. If the response rate is too low, during shear deformation, the reconstruction of the chain-like structures in time may be hindered. If the response time value is too high, the repulsions between the particles dominate, resulting in a decrease in the stability of the chain-like structure arisen. The dielectric spectra of ER fluids can usually be explained by the Cole–Cole equation and the Havriliak–Negami model [5].

3. Polyaniline-based MR fluids

MR fluids whose rheological properties tuned reversibly and rapidly (in a fraction of millisecond) from a liquid to a nearly solid state under the presence of the *H* are composed of dispersing micron-sized, soft-magnetic particles (up to 50 vol%) in a nonmagnetizable carrier liquid. Soft-magnetic particles used as a dispersed phase for MR fluids generally possess inevitable sedimentation problems due to the large density mismatch between the magnetic particles and the carrier fluid, which restricts further MR applications. Additionally, corrosion, low reversibility, and remnant magnetization of the dispersed particles are other impediments to be considered. Most common approaches to reduce density and prevent magnetic particle aggregation are application of core/shell structures with the aid of polymers and addition of surfactants or submicron-sized additives such as MWCNT, graphite nanotubes, fumed silica, and organo-clay [6]. Fabrication of core/shell structure is taken more attention compared to using additives to improve the dispersion stability of MR fluids since

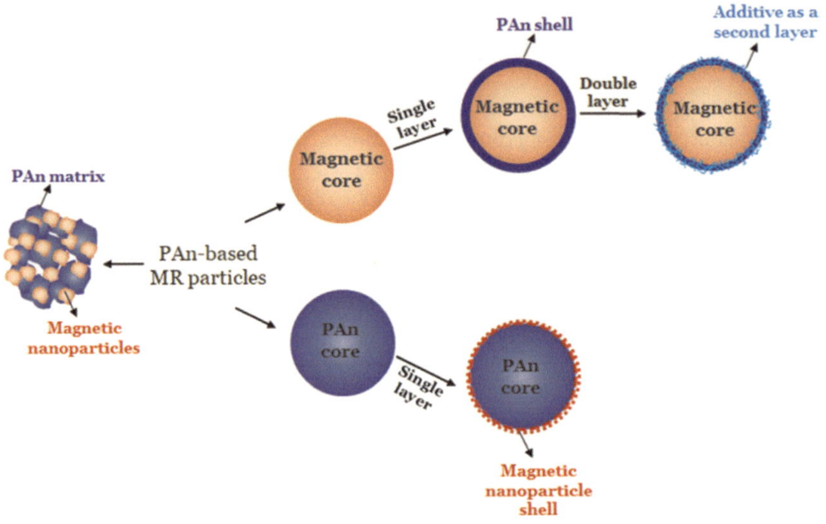

Figure 3.
Synthetic approaches for the preparation of PAn-based MR particles.

introducing additives requires further other parameters to be considered such as size, morphology, and affinity of the additive. Also, additives may interfere with the induced chain structure of the magnetic particles in H [114]. For core/shell structures, if the core is a magnetic particle, the coating shell will provide chemical and oxidation stability. On the other hand, the ultimate disintegration in shearing can be obtained when the magnetic particle is a shell and thus the interaction between magnetic shell and nonmagnetic core should be strong. Nevertheless, both approaches improve the dispersion stability [115]. As a polymer, PAn was attractive recently for enhancing the dispersion stability and obtaining dual ER/MR response [116]. Various synthetic approaches for the preparation of PAn-based MR particles are displayed in **Figure 3**. In this section, the use of PAn in MR fluids in recent studies is summarized and the comparison of various properties of PAn-based materials used for MR fluids is tabulated in **Table 1**.

3.1 Development and diversification of polyaniline-based MR fluids

Various soft-magnetic particles such as Fe_3O_4, carbonyl iron (CI), CoNi, zinc-ferrite ($ZnFe_2O_4$), and $MnFe_2O_4$ have been used as a dispersed phase for MR fluids. Their organic/inorganic composites have attracted considerable interest in terms of increasing dispersion stability of fluid, protecting magnetic particles against corrosion, and improving reversible behavior in field on/off states while maintaining intrinsic MR properties of soft-magnetic particles. However, fabricating hybrid structures have their pros and cons. Hybrid structures show slightly inferior magnetic properties compared to bare soft-magnetic particles as the aforementioned improvements are gained.

PAn having fibrous structure was coated with Fe_3O_4 by an *in-situ* self-assembly method and much lower magnetic properties were observed for the magnetic composite compared to that of pure Fe_3O_4 due to high percent composition of PAn. In a study, a magnetic composite consisting of Fe_3O_4 nanoparticles inserted in PAn matrix was fabricated and much lower magnetic properties were observed for the magnetic

Material	Magnetic particle (its M_s)	Additive	Morphology	Synthesis method	Density (g/cm^3)	M_s	MR fluid concentration	Optimum H	Yield stress at optimum H	Index α	Dual response	Ref
PAn@Fe_3O_4	Fe_3O_4 (70 emu/g)	—	Nano-sized Fe_3O_4-coated PAn tubes, approximately 148 nm in diameter and several μm in length	Self-assembly method	—	39.4 emu/g	2 vol% in SO	342 kA/m	Approximately 60 Pa	1.5	—	[117]
PAn/nano-sized Fe_3O_4 composites	Fe_3O_4 (60 emu/g)	—	Nano-sized Fe_3O_4 particles embedded into the PAn matrix	Chemical reaction method	—	18 emu/g	30 vol% in SO	343 kA/m	—	—	—	[118]
MWCNT/PAn/CI	CI	—	Micron-sized spherical, MWCNT wrapped PAn coated CI particles with rough surface	Two-step process: i. Conventional dispersion polymerization, ii. Solvent casting method	6.70	—	—	343 kA/m	9.497 kPa	1.79	—	[114]
PAn/CI	CI	0.5 wt% of nano-silica	Micron-sized spherical core–shell composite particles with a CI core and PAn shell	Coating of CI surface using a PAn colloidal dispersion prepared by emulsion polymerization	—	—	80 wt% in SO	307 mT	Approximately 8 kPa	1.5-2	—	[115]
MWCNT/PAn/CI	CI (188 Am^2/kg)	—	MWCNT nest on the PAn-coated CI core/shell spherical particles	Dispersion polymerization followed by facile solvent casting	6.70	161 Am^2/kg	20 vol% in SO	343 kA/m	10 kPa	2 at low fields and 1.5 at high fields	—	[116]

Material	Magnetic particle (its M_s)	Additive	Morphology	Synthesis method	Density (g/cm^3)	M_s	MR fluid concentration	Optimum H	Yield stress at optimum H	Index α	Dual response	Ref
CI/PAn/Fe	CI microparticles and Fe nanoparticles	PAn/Fe	PAn/Fe composite nanofibers approximately 80–150 nm in diameter with rough surface used as additive for microspherical CI	Two-step process for additive: i. Rapid mixing, oxidative polymerization, ii. Reduction of the polymerization by-products $FeCl_2/FeCl_3$ by $NaBH_4$ solution	—	(23.1 emu/g for additive PAn/Fe)	70 wt% CI and 0.1 wt% PAn/Fe nanofibers	343 kA/m	10^4 Pa	2 at low fields and 1.5 at high fields	—	[119]
CI/PAn	CI (204 emu/g)	—	CI core and PAn shell with polydisperse size distribution and rough surface	Dispersion polymerization	7.05	185 emu/g	20 vol% in SO	342 kA/m	12.5 kPa		—	[120]
CI/PAn	CI (195 emu/g)	—	Microspherical CI/PAn particles with rough surfaces and some aggregated particles	*In-situ* chemical oxidation polymerization method	4.21	136 emu/g	10 vol% in SO	343 kA/m	401 Pa (Herschel-Bulkley model)	2.0	—	[121]
PAn/ $ZnFe_2O_4$	$ZnFe_2O_4$ (91 emu/g)	—	Raspberry-like core–shell composite consisting of PAn core and $ZnFe_2O_4$ shell	Pickering emulsion polymerization	5.4	73.7 emu/g	5 vol% in SO	171 kA/m	Approximately 140 Pa	1.0	Dual	[122]

DOI: http://dx.doi.org/10.5772/intechopen.1002277

Material	Magnetic particle (its M_s)	Additive	Morphology	Synthesis method	Density (g/cm^3)	M_s	MR fluid concentration	Optimum H	Yield stress at optimum H	Index α	Dual response	Ref
$ZnFe_2O_4$/PAn	$ZnFe_2O_4$ (73.67 emu/g)	—	$ZnFe_2O_4$ core with approximately 340 nm size coated with approximately 30 nm PAn shell	*In-situ* polymerization method	2.4	43.93 emu/g	5 vol% in SO	274 kA/m	48.66 Pa	1.0	Dual	[123]
$ZnFe_2O_4$/Poly (N-methyl aniline)	$ZnFe_2O_4$ (73.67 emu/g)	—	Microspherical $ZnFe_2O_4$ core coated with poly(N-methyl aniline), approximately 428 nm particle size with rough surface	*In-situ* chemical oxidation polymerization	2.1	39.93 emu/g	5 vol% in SO	171 kA/m	47.75 Pa	—	Dual	[124]
$ZnFe_2O_4$/Poly (diphenylamine)	$ZnFe_2O_4$ (83.4 emu/g)	—	Core/shell particles with diameter of approximately 300 nm and rough surface	Radical polymerization of polymer through surface-modified $ZnFe_2O_4$ by a grafting agent	2.7	40.8 emu/g	5 vol% in SO	205 kA/m	Approximately 10^2 Pa	0.5	Dual	[125]
Poly(N-methylaniline)/ Fe_3O_4	Fe_3O_4 nanoparticles	—	Poly(N-methylaniline) microsphere coated with nano-sized Fe_3O_4 shell	Chemical co-precipitation method of Fe_3O_4 in the presence of poly (N-methyl aniline)	1.97	32.1 emu/g	10 vol% in SO	257 kA/m	177.4 Pa (Herschel-Bulkley model)	1.0	Dual	[126]

Material	Magnetic particle (its M_s)	Additive	Morphology	Synthesis method	Density (g/cm^3)	M_s	MR fluid concentration	Optimum H	Yield stress at optimum H	Index α	Dual response	Ref
$MnFe_2O_4$/PAn	$MnFe_2O_4$ (49.97 emu/g)	—	$MnFe_2O_4$/PAn particles with approximately 450 nm size, and approximately 25 nm coating thickness	*In-situ* chemical oxidation polymerization	1.89	27.46 emu/g	5 vol% in SO	171 kA/m	Approximately 20 Pa	1.0	Dual	[127]

Table 1.
The comparison of various properties of PAn-based materials used for MR fluids.

composite compared to that of pure Fe_3O_4 due to the high percent composition of PAn in the composite [118]. On the other hand, when a hierarchically structured composite was prepared by the deposition of Fe_3O_4 on fibrous structured PAn, fibrous PAn coated by Fe_3O_4 exhibited greater magnetic properties than that of particulate Fe_3O_4 embedded in PAn matrix [117]. The Fe_3O_4 particles have also been considered to fabricate a dual stimulus-responsive material to the E and the H. Fe_3O_4 coated monodispersed poly(N-methylaniline) microspheres were prepared with proper conductivity without undergoing dedoping and good magnetic susceptibility for ER and MR performance, respectively [126].

CI particles obtained from the decomposition of iron pentacarbonyl are most used for MR fluids due to their large saturation magnetization, small coercivity magnetization, and appropriate size. However, CI-based MR fluids generally have serious sedimentation problems due to the large density mismatch between the CI particles and the carrier fluid. One of the strategies to reduce the density or prevent particle aggregation is polymer coating on the particle surface. PAn coating has been introduced to reduce the density or prevent CI particle aggregation. Core-shell structured CI-PAn composite particles, prepared by dispersing CI particles in pre-synthesized PAn colloidal dispersion in chloroform, were reported to have similar magnetic properties with CI particles whereas enhanced dispersion stability, decreased field off viscosity and increased MR effect were observed compared to bare CI MR fluid [115]. To increase interfacial interaction and affinity between magnetic particles and PAn, Cl surface can be modified by suitable grafting agents. In a study, dopamine-attached CI particles were coated by PAn by dispersion polymerization process and the formed homogenous rough surface and low density of the core/shell PAn/CI particles resulted in improved dispersion stability [120]. For the same purposes, in another study, *p*-toluenesulfonic acid monohydrate was used as a surface modifier and showed the increase in chemical affinity between CI and polymer layer through hydrogen bonding and electrostatic interactions, resulting in a better core-shell morphology and improved dispersion stability, compared to the bare CI due to decreasing the density mismatch between composite particle and carrier oil [121]. Another strategy to sustain the dispersion stability was introducing carboxylic acid-functionalized MWCNT as a second layer on the PAn-coated CI core/shell particles by a two-step coating process [116]. Instead of being a shell of Cl core, iron nanoparticle-supported PAn nanofibers were used as an additive for the preparation of CI-based MR fluids and both MR properties and the dispersion stability of the MR fluid were reported to be improved [119].

Magnetic $ZnFe_2O_4$ particles coated with conducting PAn particles with raspberry-like core/shell morphology synthesized by Pickering emulsion polymerization were reported to exhibit dual ER/MR active material [122]. In another study, rough surfaces were observed when $ZnFe_2O_4$ was coated with dedoped-PAn with dual-field responsive properties [123]. Also, PAn derivatives poly(N-methyl aniline) [124] and poly (diphenylamine) [125] were also reported to be as a shell layer on magnetic $ZnFe_2O_4$ particles without any requirement of an additional de-doping process. PAn-coated spherical $MnFe_2O_4$ nanoparticles were reported with dual characteristics of ER/MR fluids [127].

3.2 MR characteristics of polyaniline-based materials

MR fluids often show a Newtonian fluid behavior, in which shear stress increases linearly with shear rate without the H. However, when a H is applied to MR fluids, the dispersed particles will get aligned along the direction of magnetic field forming fibrous structure, which leads to a solid-like state. Hence, a yield stress is necessary to make the fluid flow again. Above the yield stress, the fibrous structure breaks and yields fluidity, which means the MR fluids show a fluid-like behavior. In general, the yield stress depends on the volume fraction and the strength of H as well as the intrinsic properties of MR materials. The properties of MR fluids under the H have similarities to that of ER fluids under an electric field in terms of steady shear and oscillatory tests. However, the Herschel-Bulkley model and Bingham model are generally used to explain the flow curve of an MR fluid. The τ_y dependent on the magnetic field strength can also be expressed as $\tau_y \propto H^{\alpha}$ [128]. Instead of dielectric properties, the saturation magnetization (M_s) of a material is an important parameter predicting the MR performance that can be achieved in an MR fluid. Generally, a larger M_s value indicates higher MR performance of the MR fluid at the same particle concentration. Furthermore, the hysteresis of magnetic particles has a direct influence on the recovery of an MR fluid upon the removal of the H. If there is no hysteresis loop, the MR fluid exhibits soft magnetic properties and thus the magnetostatic interactions between particles in the carrier fluid form rapidly upon application of the H resulting in solid-like state and disappearing when the field is removed showing liquid-like state.

4. Conclusions

This chapter presents an overview of the status and recent studies in PAn-based materials used in electro/magneto-responsive smart fluids. Electro/magneto-responsive smart fluids have tremendous applications such as shock absorbers, robotics, clutches, valves, dampers, and microfluidics due to their possessing an ability to change their rheological properties reversibly and promptly under externally applied electric or magnetic fields in a controlled manner. The electro-responsive smart fluid, namely, ER fluids mainly consists of polarizable particles dispersed in nonconductive liquid dispersant and furthermore, various polar additives may be inserted to increase polarizability or dispersion stability of the ER fluid. But, instead of incorporating additives due to their drawbacks such as leading electrical breakdown and device corrosion, inherently anhydrous polarizable particles are desired to be used as an ER active material. Thus, inherently polarizable PAn has been one of the most promising and extensively studied conducting polymers in the research of ER materials for many years due to its tunable conductivity, ease of synthesis, adequate chemical and thermal stability, noncorrosiveness, and less friction than pure inorganic compounds. The morphology, size, shape, surface characteristics, dedoping level, and incorporation of other compounds with PAn have a significant influence on the ER properties. In this chapter, PAn-based ER fluids with various compositions are categorized in terms of constituent components and discussed in detail. Various inorganic compounds, insulating polymers, PILs, and carbon-based nanomaterials have been hybridized with PAn and PAn derivatives to improve the ER performance and the dispersion stability.

The magnetic field analog of ER fluids, MR fluids are basically made up of soft magnetizable dispersed particles, carrier fluid, and additives. The researchers have focused on enhancing the dispersion stability of soft-magnetizable dispersed particles in MR fluid, preventing particle corrosion, and improving the overall MR effect by reduction in density mismatch between carrier fluid and dispersed particles, the addition of various additives and particle coating with inorganic, organic or polymeric components. Even though PAn has no soft magnetic properties, PAn and its derivatives have also been used for enhancing the dispersion stability of MR fluids and obtaining dual ER/MR response. For that reason, there have been fewer studies on the use of PAn in MR fluids compared to its use in the ER fluids. The literature studies have indicated that while introducing PAn in MR materials enhanced the sedimentation stability, the magnetic saturation of the resulting structure was decreased. Therefore, there are still major opportunities for the development of PAn with low magnetic masking property in the future research.

According to the recent studies reported so far, it has been concluded that owing high surface area, good wettability, rough surface compared to smooth surface, hierarchical anisotropic structures compared to isotropic ones, suitable conductivity, and reduced density mismatch with carrier fluid led to enhanced ER/MR performances of PAn-based stimuli responsive fluids.

Author details

Ozlem Erol
Chemistry Department, Science Faculty, Gazi University, Ankara, Turkey

*Address all correspondence to: oerol@gazi.edu.tr

References

[1] Hao T. Electrorheological fluids. Advanced Materials. 2001;**13**(24): 1847-1857

[2] de Vicente J, Klingenberg DJ, Hidalgo-Alvarez R. Magnetorheological fluids: A review. Soft Matter. 2011;**7**(8): 3701-3710

[3] Winslow WM. Induced fibration of suspensions. Journal of Applied Physics. 1949;**20**(12):1137-1140

[4] Hao T. Electrorheological suspensions. Advances in Colloid and Interface Science. 2002;**97**(1):1-35

[5] Dong YZ, Kim HM, Choi HJ. Conducting polymer-based electro-responsive smart suspensions. Chemical Papers. 2021;**75**(10):5009-5034

[6] Kumar JS, Paul PS, Raghunathan G, Alex DG. A review of challenges and solutions in the preparation and use of magnetorheological fluids. International Journal of Mechanical and Materials Engineering. 2019;**14**(1):13

[7] Rabbani Y, Hajinajaf N, Shariaty-Niassar M. The effect of microparticles/nanoparticles surface modification on the magnetorheological fluid properties: A review. Journal of Intelligent Material Systems and Structures. 2023;**0**(0): 1045389X221147667

[8] Kuznetsov NM, Kovaleva VV, Belousov SI, Chvalun SN. Electrorheological fluids: from historical retrospective to recent trends. Materials Today Chemistry. 2022;**26**:101066

[9] Stenicka M, Pavlinek V, Saha P, Blinova NV, Stejskal J, Quadrat O. The effect of compatibility of suspension particles with the oil medium on electrorheological efficiency. Journal of Intelligent Material Systems and Structures. 2012;**23**(9):1055-1059

[10] Stěnička M, Pavlínek V, Sáha P, Blinova NV, Stejskal J, Quadrat O. Effect of hydrophilicity of polyaniline particles on their electrorheology: Steady flow and dynamic behaviour. Journal of Colloid and Interface Science. 2010;**346**(1): 236-240

[11] Xie H-Q, Guan J-G, Guo J-S. Three ways to improve electrorheological properties of polyaniline-based suspensions. Journal of Applied Polymer Science. 1997;**64**:1641-1647

[12] Sebastian J, Samuel JM. Recent advances in the applications of substituted polyanilines and their blends and composites. Polymer Bulletin. 2020; **77**(12):6641-6669

[13] Fang FF, Lee BM, Choi HJ. Electrorheologically intelligent polyaniline and its composites. Macromolecular Research. 2010;**18**(2): 99-112

[14] Kuramoto N, Takahashi Y, Nagai K, Koyama K. Electrorheological properties of poly(o-anisidine) and poly(o-anisidine)-coated silica suspensions. Reactive and Functional Polymers. 1996; **30**(1):367-373

[15] Liu J, Wen X, Liu Z, Tan Y, Yang S, Zhang P. Electrorheological performances of poly(o-toluidine) and p-toluenesulfonic acid doped poly(o-toluidine) suspensions. Colloid and Polymer Science. 2015;**293**(5):1391-1400

[16] Gercek B, Yavuz M, Yilmaz H, Sari B, Unal HI. Comparison of electrorheological properties of some polyaniline derivatives. Colloids and Surfaces A: Physicochemical and

Engineering Aspects. 2007;**299**(1): 124-132

[17] Kim JW, Jang WH, Choi HJ, Joo J. Synthesis and electrorheological characteristics of polyaniline derivatives with different substituents. Synthetic Metals. 2001;**119**(1):173-174

[18] Kim MH, Bae DH, Choi HJ, Seo Y. Synthesis of semiconducting poly (diphenylamine) particles and analysis of their electrorheological properties. Polymer. 2017;**119**:40-49

[19] Woo DJ, Suh MH, Shin ES, Lee CW, Lee SH. Electrorheological behavior of suspensions of a substituted polyaniline with long alkyl pendants. J Colloid Interface Sci. 2005; **288**(1):71-74

[20] Han WJ, Choi HJ. Synthesis of conducting polymeric nanoparticles in the presence of a polymerizable surfactant and their electrorheological response. Colloid and Polymer Science. 2019;**297**(5):781-784

[21] Choi HJ, Kim JW, To K. Electrorheological characteristics of semiconducting poly(aniline-co-o-ethoxyaniline) suspension. Polymer. 1999;**40**(8):2163-2166

[22] Trlica J, Sáha P, Quadrat O, Stejskal J. Electrical and electrorheological behavior of poly (aniline-co-1,4-phenylenediamine) suspensions. European Polymer Journal. 2000;**36**(11):2313-2319

[23] Cho MS, Kim TW, Choi HJ, Jhon MS. N-substituted copolyaniline for electrorheological material. Journal of Materials Science Letters. 1997;**16**: 672-673

[24] Cho CH, Choi HJ, Kim JW, Jhon MS. Synthesis and electrorheology of aniline/ pyrrole copolymer. Journal of Materials Science. 2004;**39**(5):1883-1885

[25] Mrlik M, Sedlacik M, Pavlinek V, Bober P, Trchová M, Stejskal J, et al. Electrorheology of aniline oligomers. Colloid and Polymer Science. 2013; **291**(9):2079-2086

[26] Sung BH, Choi US, Jang HG, Park YS. Novel approach to enhance the dispersion stability of ER fluids based on hollow polyaniline sphere particle. Colloids and Surfaces A: Physicochemical and Engineering Aspects. 2006;**274**(1):37-42

[27] Yin J, Zhao X, Xia X, Xiang L, Qiao Y. Electrorheological fluids based on nano-fibrous polyaniline. Polymer. 2008;**49**(20):4413-4419

[28] Fang FF, Dong Y-Z, Choi HJ. Effect of oxidants on morphology of interfacial polymerized polyaniline nanofibers and their electrorheological response. Polymer. 2018;**158**:176-182

[29] Yin J, Xia X, Xiang L, Zhao X. Conductivity and polarization of carbonaceous nanotubes derived from polyaniline nanotubes and their electrorheology when dispersed in silicone oil. Carbon. 2010;**48**(10): 2958-2967

[30] Lee BM, Kim JE, Fang FF, Choi HJ, Feller J-F. Rectangular-shaped polyaniline tubes covered with nanorods and their electrorheology. Macromolecular Chemistry and Physics. 2011;**212**(21):2300-2307

[31] Kim MJ, Liu YD, Choi HJ. Urchin-like polyaniline microspheres fabricated from self-assembly of polyaniline nanowires and their electro-responsive characteristics. Chemical Engineering Journal. 2014;**235**:186-190

[32] Wen Q, He K, Wang C, Wang B, Yu S, Hao C, et al. Clip-like polyaniline nanofibers synthesized by an insitu chemical oxidative polymerization and its strong electrorheological behavior. Synthetic Metals. 2018;**239**:1-12

[33] Yin J, Xia X, Xiang L, Qiao Y, Zhao X. The electrorheological effect of polyaniline nanofiber, nanoparticle and microparticle suspensions. Smart Materials and Structures. 2009;**18**(9): 095007

[34] Lin DD, Zhang ZJ, Zhao BY, Chen LS, Hu K. Rapid synthesis of porous polyaniline and its application in electrorheological fluid. Smart Materials and Structures. 2006;**15**(6):1641

[35] Choi HJ, Cho MS, To K. Electrorheological and dielectric characteristics of semiconductive polyaniline-silicone oil suspensions. Physica A: Statistical Mechanics and its Applications. 1998;**254**(1):272-279

[36] Liu YD, Choi HJ. Electrorheological response of polyaniline and its hybrids. Chemical Papers. 2013;**67**(8):849-859

[37] Ji L, Zhang J. Synthesis, characterization and electrorheological properties of polyaniline/titanate core-shell composite. Journal of Macromolecular Science, Part A. 2009; **46**(7):688-693

[38] Cheng Q, Pavlinek V, He Y, Li C, Saha P. Electrorheological characteristics of polyaniline/titanate composite nanotube suspensions. Colloid and Polymer Science. 2009;**287**(4):435-441

[39] Yin J, Xia X, Wang X, Zhao X. The electrorheological effect and dielectric properties of suspensions containing polyaniline@titania nanocable-like particles. Soft Matter. 2011;**7**(22): 10978-10986

[40] Wang B, Liu C, Yin Y, Yu S, Chen K, Liu P, et al. Double template assisting synthesized core–shell structured titania/polyaniline nanocomposite and its smart electrorheological response. Composites Science and Technology. 2013;**86**:89-100

[41] Lee IS, Lee JY, Sung JH, Choi HJ. Synthesis and electrorheological characteristics of polyaniline-titanium dioxide hybrid suspension. Synthetic Metals. 2005;**152**(1):173-176

[42] Tang J, Wen X, Liu Z, Wang J, Zhang P. Synthesis and electrorheological performances of 2D PANI/TiO_2 nanosheets. Colloids and Surfaces A: Physicochemical and Engineering Aspects. 2018;**552**:24-31

[43] Lee S, Noh J, Jekal S, Kim J, Oh W-C, Sim H-S, et al. Hollow TiO_2 nanoparticles capped with polarizability-tunable conducting polymers for improved electrorheological activity. Nanomaterials. 2022;**12**(19):3521

[44] Wang B, Liu C, Yin Y, Tian X, Yu S, Chen K, et al. The electrorheological properties of polyaniline nanofiber/kaolinite hybrid nanocomposite. Journal of Applied Polymer Science. 2013; **130**(2):1104-1113

[45] Lengálová A, Vr P, Sáha P, Stejskal J, Kitano T, Quadrat O. The effect of dielectric properties on the electrorheology of suspensions of silica particles coated with polyaniline. Physica A: Statistical Mechanics and its Applications. 2003;**321**(3):411-424

[46] Liu YD, Zhang WL, Choi HJ. Pickering emulsion polymerization of core–shell-structured polyaniline@SiO_2 nanoparticles and their electrorheological response. Colloid and Polymer Science. 2012;**290**(9):855-860

[47] Jang HS, Kwon SH, Lee JH, Choi HJ. Facile fabrication of core-shell typed silica/poly(diphenylamine) composite microparticles and their electro-response. Polymer. 2019;**182**:121851

[48] Zhang WL, Choi HJ. Fabrication of semiconducting polyaniline-wrapped halloysite nanotube composite and its electrorheology. Colloid and Polymer Science. 2012;**290**(17):1743-1748

[49] Marins JA, Giulieri F, Soares BG, Bossis G. Hybrid polyaniline-coated sepiolite nanofibers for electrorheological fluid applications. Synthetic Metals. 2013;**185-186**:9-16

[50] Jang DS, Choi HJ. Conducting polyaniline-wrapped sepiolite composite and its stimuli-response under applied electric fields. Colloids and Surfaces A: Physicochemical and Engineering Aspects. 2015;**469**:20-28

[51] Tian X, He K, Wang B, Yu S, Hao C, Chen K, et al. Flower-like Fe_2O_3/polyaniline core/shell nanocomposite and its electroheological properties. Colloids and Surfaces A: Physicochemical and Engineering Aspects. 2016;**498**:185-193

[52] Zhu X, Zhao Q, Zhang T, Pang X. Electrorheological response of novel polyaniline-Fe_2O_3 nanocomposite particles. Polymer-Plastics Technology and Materials. 2019;**58**(5):573-577

[53] Dong YZ, Choi HJ. Electrorheological characteristics of poly(diphenylamine)/magnetite composite-based suspension. Materials. 2019;**12**(18):2911

[54] Lu Q, Lee JH, Lee JH, Choi HJ. Magnetite/poly(ortho-anisidine) composite particles and their electrorheological response. Materials. 2021;**14**(11):2900

[55] Chae HS, Zhang WL, Piao SH, Choi HJ. Synthesized palygorskite/polyaniline nanocomposite particles by oxidative polymerization and their electrorheology. Applied Clay Science. 2015;**107**:165-172

[56] Fei Fang F, Hye Kim J, Jin Choi H, Seo Y. Organic/inorganic hybrid of polyaniline/$BaTiO_3$ composites and their electrorheological and dielectric characteristics. Journal of Applied Polymer Science. 2007;**105**(4):1853-1860

[57] Yan H, Liao ZJ, Zhu X, Wang XM, Chen Y, Zhang B, et al. Synthesis and electrorheological properties of polyaniline-coated barium titanate composite particles. Journal of Applied Polymer Science. 2008;**107**(3):1960-1966

[58] Linzhi LI, Shujuan GAO. Polyaniline (PANI) and $BaTiO_3$ composite nanotube with high suspension performance in electrorheological fluid. Materials Today Communications. 2020;**24**:100993

[59] Wang B, Yin Y, Liu C, Yu S, Chen K. Synthesis and characterization of clay/polyaniline nanofiber hybrids. Journal of Applied Polymer Science. 2013;**128**(2):1304-1312

[60] Kim JW, Kim SG, Choi HJ, Jhon MS. Synthesis and electrorheological properties of polyaniline-Na^+-montmorillonite suspensions. Macromolecular Rapid Communications. 1999;**20**(8):450-452

[61] Lu J, Zhao X. Electrorheological properties of a polyaniline–montmorillonite clay nanocomposite suspension. Journal of Materials Chemistry. 2002;**12**(9):2603-2605

[62] Song DH, Lee HM, Lee K-H, Choi HJ. Intercalated conducting polyaniline–clay nanocomposites and

their electrical characteristics. Journal of Physics and Chemistry of Solids. 2008; **69**(5):1383-1385

[63] Lim YT, Park JH, Park OO. Improved electrorheological effect in polyaniline nanocomposite suspensions. Journal of Colloid and Interface Science. 2002;**245**(1):198-203

[64] Fang FF, Liu YD, Choi HJ. Synthesis and electrorheological characteristics of polyaniline/organoclay nanoparticles via Pickering emulsion polymerization. Smart Materials and Structures. 2010; **19**(12):124002

[65] Sung JH, Choi HJ. Effect of pH on physical characteristics of conducting poly(o-ethoxyaniline) nanocomposites. Journal of Macromolecular Science, Part B. 2005;**44**(3):365-375

[66] Jun CS, Sim B, Choi HJ. Fabrication of electric-stimuli responsive polyaniline/laponite composite and its viscoelastic and dielectric characteristics. Colloids and Surfaces A: Physicochemical and Engineering Aspects. 2015;**482**:670-677

[67] Cho MS, Choi HJ, Kim KY, Ahn WS. Synthesis and characterization of polyaniline/mesoporous SBA-15 nanocomposite. Macromolecular Rapid Communications. 2002;**23**(12):713-716

[68] Cho MS, Choi HJ, Ahn W-S. Enhanced electrorheology of conducting polyaniline confined in MCM-41 channels. Langmuir. 2004;**20**(1):202-207

[69] Fang FF, Choi HJ, Ahn WS. Electroactive response of mesoporous silica and its nanocomposites with conducting polymers. Composites Science and Technology. 2009;**69**(13):2088-2092

[70] Noh J, Yoon C-M, Jang J. Enhanced electrorheological activity of polyaniline coated mesoporous silica with high aspect ratio. Journal of Colloid and Interface Science. 2016;**470**:237-244

[71] Wei C, Zhu Y, Yang X, Li C. One-pot synthesis of polyaniline-doped in mesoporous TiO_2 and its electrorheological behavior. Materials Science and Engineering: B. 2007;**137**(1): 213-216

[72] Tian X, He K, Wang C, Wen Q, Wang B, Yu S, et al. Preparation and electrorheological behavior of anisotropic titanium oxide/polyaniline core/shell nanocomposite. Composites Science and Technology. 2016;**137**: 118-129

[73] Han WJ, Piao SH, Choi HJ. Synthesis and electrorheological characteristics of polyaniline@attapulgite nanoparticles via Pickering emulsion polymerization. Materials Letters. 2017;**204**:42-44

[74] Chattopadhyay A, Rani P, Srivastava R, Dhar P. Electro-elastoviscous response of polyaniline functionalized nano-porous zeolite based colloidal dispersions. Journal of Colloid and Interface Science. 2018;**519**:242-254

[75] Wen Q, Ma L, Wang C, Wang B, Han R, Hao C, et al. Preparation of core–shell structured metal–organic framework@PANI nanocomposite and its electrorheological properties. RSC Advances. 2019;**9**(25):14520-14530

[76] Zhang WL, Jiang D, Wang X, Hao BN, Liu YD, Liu J. Growth of polyaniline nanoneedles on MoS_2 nanosheets, tunable electroresponse, and electromagnetic wave attenuation analysis. The Journal of Physical Chemistry C. 2017;**121**(9):4989-4998

[77] Li C, Chen Y, Wang L, Wang Z, Lin Y, Xiong K, et al. Electrorheological response behavior of PANI@MoS_2 core–

shell nanocomposites. Advanced Engineering Materials. 2023;**25**:2300029

[78] Stejskal J, Mrlík M, Plachý T, Trchová M, Kovářová J, Li Y. Molybdenum and tungsten disulfides surface-modified with a conducting polymer, polyaniline, for application in electrorheology. Reactive and Functional Polymers. 2017;**120**:30-37

[79] Erol O, Karakisla M, Unal HI, Sacak M. Electrorheological properties of polyaniline/K-feldspar conducting composite. Journal of Composite Materials. 2012;**46**(11):1295-1304

[80] Yavuz M, Gok A, Sen S, Unal HI. Electrorheological properties of polyaniline/red mud composite. International Journal of Polymer Analysis and Characterization. 2008;**13**(1):9-24

[81] Lengálová A, Pavlínek Vr, Sáha P, Stejskal J, Quadrat O. Electrorheology of polyaniline-coated inorganic particles in silicone oil. Journal of Colloid and Interface Science. 2003;**258**(1):174-178

[82] Munteanu L, Munteanu A, Sedlacik M, Kutalkova E, Kohl M, Kalendova A. Zinc ferrite/polyaniline composite particles: Pigment applicable as electro-active paint. J Ind Eng Chem. 2022;**115**:440-448

[83] Zhang WL, Liu J, Choi HJ. Graphene and graphene oxide composites and their electrorheological applications. Journal of Nanomaterials. 2015;**2015**:574637

[84] Lu Q, Jang HS, Han WJ, Lee JH, Choi HJ. Stimuli-responsive graphene oxide-polymer nanocomposites. Macromolecular Research. 2019;**27**(11): 1061-1070

[85] Park SJ, Park SY, Cho MS, Choi HJ, Jhon MS. Synthesis and electrorhelogy of multi-walled carbon nanotube/ polyaniline nanoparticles. Synthetic Metals. 2005;**152**(1):337-340

[86] Sim B, Zhang WL, Choi HJ. Graphene oxide/poly(2-methylaniline) composite particle suspension and its electro-response. Materials Chemistry and Physics. 2015;**153**:443-449

[87] Zhang K, Li H, Dong YZ, Zhang H, Zhao W, Zhao S, et al. Jellyfish-shaped p-phenylenediamine functionalized graphene oxide-g-polyaniline fibers and their electrorheology. Polymer. 2019; **168**:29-35

[88] Gao CY, Kim MH, Jin H-J, Choi HJ. Synthesis and electrorheological response of graphene oxide/ polydiphenylamine microsheet composite particles. Polymers. 2020; **12**(9):1984

[89] Yin J, Wang X, Chang R, Zhao X. Polyaniline decorated graphene sheet suspension with enhanced electrorheology. Soft Matter. 2012;**8**(2): 294-297

[90] Yuan J, Wang Y, Xiang L, Zhao X, Yin J. Understanding the enhanced electrorheological effect of reduced graphene oxide-supported polyaniline dielectric nanoplates by a comparative study with graphene oxide as the support core. IET Nanodielectrics. 2021;**4**(3): 143-154

[91] Sedlacik M, Almajdalawi S, Mrlik M, Pavlinek V, Saha P, Stejskal J. Viscoelastic properties of electrorheological suspensions of core-shell (carbon/polyaniline) particles in silicone oil. Journal of Physics: Conference Series. 2013;**412**(1):012006

[92] Yin J, Shui Y, Chang R, Zhao X. Graphene-supported carbonaceous dielectric sheets and their

electrorheology. Carbon. 2012;**50**(14): 5247-5255

[93] Sedlacik M, Pavlinek V, Mrlik M, Morávková Z, Hajná M, Trchová M, et al. Electrorheology of polyaniline, carbonized polyaniline, and their core–shell composites. Materials Letters. 2013; **101**:90-92

[94] Abu Hassan Shaari H, Ramli MM, Mohtar MN, Abdul Rahman N, Ahmad A. Synthesis and conductivity studies of poly(methyl methacrylate) (PMMA) by co-polymerization and blending with polyaniline (PANi). Polymers. 2021;**13**(12):1939

[95] Kim MH, Choi HJ. Core–shell structured semiconducting poly (diphenylamine)-coated polystyrene microspheres and their electrorheology. Polymer. 2017;**131**:120-131

[96] Sung SK, Hyoung JC. Poly(2,5-dimethoxyaniline) coated polystyrene microspheres and their electrorheological characteristics. Colloids and Surfaces A: Physicochemical and Engineering Aspects. 2020;**605**:125375

[97] Park SY, Cho MS, Kim CA, Choi HJ, Jhon MS. Polyaniline microsphere encapsulated by poly(methyl methacrylate) and investigation of its electrorheological properties. Colloid and Polymer Science. 2003;**282**(2): 198-202

[98] Liu YD, Fang FF, Choi HJ. Core–shell structured semiconducting PMMA/polyaniline snowman-like anisotropic microparticles and their electrorheology. Langmuir. 2010;**26**(15): 12849-12854

[99] Lee IS, Cho MS, Choi HJ. Preparation of polyaniline coated poly (methyl methacrylate) microsphere by graft polymerization and its electrorheology. Polymer. 2005;**46**(4): 1317-1321

[100] Jun J-B, Kim J-W, Suh K-D. Monodisperse micron-sized polyaniline composite particles for electrorheological fluid material. Macromolecular Chemistry and Physics. 2002;**203**(7):1011-1017

[101] Cho MS, Cho YH, Choi HJ, Jhon MS. Synthesis and electrorheological characteristics of polyaniline-coated poly(methyl methacrylate) microsphere: Size effect. Langmuir. 2003;**19**(14):5875-5881

[102] Moon IJ, Choi HJ. Core-shell-structured copolyaniline-coated polymeric nanoparticle suspension and its viscoelastic response under various electric fields. Materials. 2015;**8**(8): 4932-4942

[103] Jun CS, Kwon SH, Choi HJ, Seo Y. Polymeric nanoparticle-coated Pickering emulsion-synthesized conducting polyaniline hybrid particles and their electrorheological study. ACS Applied Materials & Interfaces. 2017;**9**(51): 44811-44819

[104] Dong Y, Yin J, Zhao X. Microwave-synthesized poly(ionic liquid) particles: A new material with high electrorheological activity. Journal of Materials Chemistry A. 2014;**2**(25): 9812-9819

[105] Dong Y, Liu Y, Wang B, Xiang L, Zhao X, Yin J. Influence of counterion type on dielectric and electrorheological responses of poly(ionic liquid)s. Polymer. 2017;**132**:273-285

[106] Liu Y, Zhao J, He F, Zheng C, Zhao X, Yin J. Influence of alkyl spacer length on ion transport, polarization and electro-responsive electrorheological

effect of self-crosslinked poly(ionic liquid)s. Polymer. 2019;**171**:161-172

[107] Zheng C, Dong Y, Liu Y, Zhao X, Yin J. Enhanced stimuli-responsive electrorheological property of poly(ionic liquid)s-capsulated polyaniline particles. Polymers. 2017;**9**(9):385

[108] Zheng C, Liu Y, Dong Y, He F, Zhao X, Yin J. Low-temperature interfacial polymerization and enhanced electro-responsive characteristic of poly (ionic liquid)s@polyaniline core-shell microspheres. Macromolecular Rapid Communications. 2019;**40**(17):1800351

[109] Zheng C, Lei Q, Zhao J, Zhao X, Yin J. The effect of dielectric polarization rate difference of filler and matrix on the electrorheological responses of poly (ionic liquid)/polyaniline composite particles. Polymers. 2020;**12**(3):703

[110] Li X, Yan G, Wang J, Kong W, Chang X, Zhuang Y, et al. Effect of a temperature threshold on the electrorheological performance of ionic liquid crystal polyanilines. Journal of Molecular Liquids. 2021;**326**:115299

[111] Meng F-B, Zhou N-Y, Du C, Gao H-M, He X-Z. Synthesis, characterization and electrorheological effect of sulfosalt-type liquid-crystalline ionomers containing polyaniline units. Journal of Applied Polymer Science. 2013;**130**(5): 3395-3403

[112] Seo YP, Choi HJ, Lee JR, Seo Y. Modeling and analysis of an electrorheological flow behavior containing semiconducting graphene oxide/polyaniline composite particles. Colloids and Surfaces A: Physicochemical and Engineering Aspects. 2014;**457**:363-367

[113] Ramli H, Zainal NFA, Hess M, Chan CH. Basic principle and good practices of rheology for polymers for teachers and beginners. Chemistry Teacher International. 2022;**4**(4): 307-326

[114] Fang FF, Choi HJ, Choi WS. Two-layer coating with polymer and carbon nanotube on magnetic carbonyl iron particle and its magnetorheology. Colloid and Polymer Science. 2010; **288**(3):359-363

[115] Sedlačík M, Pavlínek V, Sáha P, Švrčinová P, Filip P, Stejskal J. Rheological properties of magnetorheological suspensions based on core–shell structured polyaniline-coated carbonyl iron particles. Smart Materials and Structures. 2010;**19**(11): 115008

[116] Fang FF, Liu YD, Choi HJ, Seo Y. Core–shell structured carbonyl iron microspheres prepared via dual-step functionality coatings and their magnetorheological response. ACS Applied Materials & Interfaces. 2011; **3**(9):3487-3495

[117] Kwon SH, Sim B, Choi HJ. Magnetorheological characteristics of nano-sized iron oxide coated polyaniline composites. IEEE Transactions on Magnetics. 2016;**52**(7):1-4

[118] Kim JH, Fang FF, Choi HJ, Seo Y. Magnetic composites of conducting polyaniline/nano-sized magnetite and their magnetorheology. Materials Letters. 2008;**62**(17):2897-2899

[119] Piao SH, Bhaumik M, Maity A, Choi HJ. Polyaniline/Fe composite nanofiber added softmagnetic carbonyl iron microsphere suspension and its magnetorheology. Journal of Materials Chemistry C. 2015;**3**(8):1861-1868

[120] Moon IJ, Kim MW, Choi HJ, Kim N, You C-Y. Fabrication of dopamine

grafted polyaniline/carbonyl iron core-shell typed microspheres and their magnetorheology. Colloids and Surfaces A: Physicochemical and Engineering Aspects. 2016;**500**:137-145

[121] Min TH, Choi HJ, Kim N-H, Park K, You C-Y. Effects of surface treatment on magnetic carbonyl iron/polyaniline microspheres and their magnetorheological study. Colloids and Surfaces A: Physicochemical and Engineering Aspects. 2017;**531**:48-55

[122] Kim JN, Dong YZ, Choi HJ. Pickering emulsion polymerized polyaniline/zinc-ferrite composite particles and their dual electrorheological and magnetorheological responses. ACS Omega. 2020;**5**(13):7675-7682

[123] Kim HM, Kang SH, Choi HJ. Polyaniline coated $ZnFe_2O_4$ microsphere and its electrorheological and magnetorheological response. Colloids and Surfaces A: Physicochemical and Engineering Aspects. 2021;**626**:127079

[124] Kim HM, Jeong JY, Kang SH, Jin H-J, Choi HJ. Dual electrorheological and magnetorheological behaviors of poly (N-methyl aniline) coated $ZnFe_2O_4$ composite particles. Materials. 2022; **15**(7):2677

[125] Kim TH, Choi HJ. Fabrication and shear response of conducting polymer-coated zinc ferrite particles under magnetic/electric field. IEEE Transactions on Magnetics. 2022;**58**(2): 1-5

[126] Dong Y, Wang S, Choi HJ. Poly(N-methylaniline)/magnetite microsphere and its electrical and magnetic dual responses. Polymer. 2022;**240**:124492

[127] Jeong JY, Kim S, Baek E, You CY, Choi HJ. Suspension rheology of polyaniline coated manganese ferrite particles under electric/magnetic fields. Colloids and Surfaces A: Physicochemical and Engineering Aspects. 2023;**656**:130438

[128] Dong YZ, Choi K, Kwon SH, Nam J-D, Choi HJ. Nanoparticles functionalized by conducting polymers and their electrorheological and magnetorheological applications. Polymers. 2020;**12**(1):204

Chapter 4

Polyaniline for Smart Textile Applications

Lihi Abilevitch, Limor Mizrahi, Gali Cohen, Shmuel Kenig and Elizabeth Amir

Abstract

With the development of smart and functional textiles, electro-conductive fabrics based on polyaniline have attracted much attention due to its unique chemical structure, ease of preparation, flexibility, stability, excellent electrical conductivity, and sensing properties. As a result, polyaniline-based fabrics are widely used in various applications, including electromagnetic shielding, electronics, sensing, monitoring, and biomedicine. This chapter reviews the state-of-the-art technologies for fabricating polyaniline-coated woven, non-woven, and knitted fabrics based on natural and synthetic polymers, describing the fabrication methods, characterization techniques, and applications.

Keywords: polyaniline-coated fabric, polyaniline, antibacterial textile, electro-conductive fabric, mechanical sensor, ammonia gas sensor, pH sensor

1. Introduction

Textiles are used daily in several applications, such as apparel, household textiles, furniture, medical and protective equipment, buildings, and vehicles. Textiles enable products' functionalities and performance as well as esthetics and comfort. The contemporary textile industry is continuously challenged with innovative applications based on new technologies. Therefore, "smart textiles," "E-textiles," "functional textiles," and "wearable electronic textiles" are among the terms used to represent the potential applications. Smart textiles have become the most promising next-generation wearable fabrics [1–7]. Smart textiles can respond to external physical and chemical stimuli, communicate by sensor technology, and adapt to changing surroundings [5, 6, 8]. The main aim behind developing smart textiles is to introduce new applications using efficient methodologies without compromising the intrinsic comfort, flexibility, and light weight of the fabric. Smart textiles have enormous potential in different applications possessing properties, including antibacterial, thermochromic (response to changes in temperature), shape-memory (response to external stimuli), hydrophobic and hydrophilic, oil-water separation, controlled release, drug delivery, photonics, sensors (response to pressure, pH, gas), antistatic, energy storage, and electro-conductive properties [1, 9–17].

Common textiles do not conduct electricity and are considered insulators. The textiles become electrically conductive by incorporating electro-conductive materials such as graphene (G), graphene oxide (GO), conductive nano, microparticles, and conjugated organic polymers on their surface for applications such as wearable electronic textiles [5, 6, 18–20]. The electrical conductivity of textiles depends on various factors, including the type and nature of the conductive coating as well as the ability to form an efficient interconnected conductive network within porous textile substrate. Incorporating conductive metallic particles often results in rigid fabrics with reduced flexibility that cannot maintain their functionality after washing. Therefore, flexible, deformable, stretchable, and durable electro-conductive threads are critical for electro-conductive textiles that serve as smart textiles for information transfer and computation capability while accommodating the movement of the human body [5, 6, 15, 21–26]. For those reasons, intrinsic conductive polymers are the most suitable candidates for replacing conductive fillers to fabricate conductive textiles.

In recent decades, intrinsically conductive polymers (ICPs) have been extensively studied due to their unique electronic and electro-optical properties, and they are an essential part of the development of smart textiles [27–29]. In 1976, Shirakawa, MacDiarmid, and Heeger made a breakthrough discovery of a thin polyacetylene film oxidized with iodine vapor resulting in electrical conductivity [30]. They were awarded the Nobel Prize in Chemistry for discovering electrical conductivity in organic-conjugated materials in 2000 [31]. Over the years, dozens of ICPs have been introduced, among them polypyrrole (PPy), polyaniline (PANI), polythiophenes (PThs), and their derivatives. Their exceptional combination of properties, including electrical conductivity, electromagnetic shielding ability, light weight, flexibility, low cost, solution processability, good adhesion to diverse substrates, and ease of preparation, makes them ideal candidates for the development of smart textiles [21, 32, 33]. PANI is one of the most applied organic-conjugated polymers on textile due to its unique chemical structure, environmental stability, ease of preparation in aqueous solution, and excellent film-forming ability [34]. The electrical conductivity usually correlates to the molecular weight of the conjugated polymer, having reported that PANI with high molecular weight achieved higher electrical conductivity than the low molecular weight derivatives [35–38]. The molecular structure of PANI consists of repeating units of the monomer aniline, as shown in **Figure 1**. The location of the N-atom between the phenyl rings enables the formation of various oxidation states, which significantly affect the physicochemical and mechanical properties of PANI [39]. There are three basic oxidation states of PANI: (a) leucoemeraldine (white/clear), (b) emeraldine (salt-green/base-blue), and (c) pernigraniline (blue/violet) [40, 41]. PANI exhibits two distinctive interconverting molecular forms in emeraldine oxidation state: salt (emeraldine salt (ES)) and base (emeraldine base (EB)). EB is a neutral form in which the amine groups are present as imines. Protonation of the imine nitrogen group with acid results in the doped ES state, in which the imines are converted into quaternary ammonium ions (**Figure 2A**) [42]. ES is the only conducting state among various forms of PANI that can be produced in an acidic solution ($pH < 3.0$) [43]. Doping with inorganic acids, such as hydrochloric or sulfuric acid, can influence the conductivity of PANI. As a result of doping, a charge transfer reaction occurs, which in turn creates active sites (polarons), adding an electron to the conduction band (n-doping) or removing an electron from the valence band (p-doping) [39]. The dopant degree can be easily changed by controlling the pH value of the dopant acid/base solution [44]. Both PANI-ES and PANI-EB display three

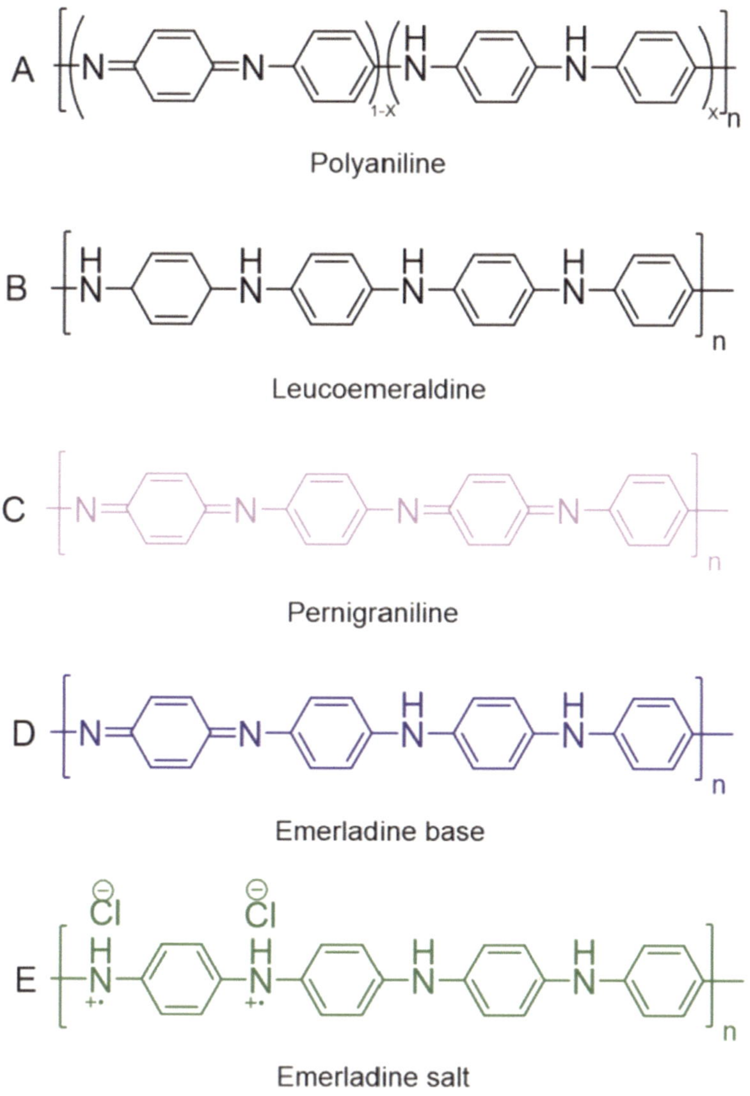

Figure 1.
Chemical structures of the different forms of polyaniline (PANI): (A) polyaniline, (B) leucoemeradine, (C) pernigraniline, (D) emeraldine base (EB), and (E) emeraldine salt (ES).

ultraviolet-visible (UV-vis) absorption peaks. The absorption peaks between 255 and 286 nm and 328–349 nm are attributed to the π-π* transitions of the conjugated benzene rings of the PANI-ES and PANI-EB backbones. The third peak of PANI-ES located at 430–439 nm is due to polaron absorption resulting from the doping process. While the third band of PANI-EB is located at 610–650 nm, indicating the conversion of the imine nitrogen atoms of the benzenoid rings to quinonoid rings [45–49]. These unique properties of polyaniline make this remarkable material a suitable candidate for the development of chemical sensors [50], electromagnetic shielding devices [51], antibacterial protection [52], and electrochromic and electrochemical devices [53, 54] for smart textile applications.

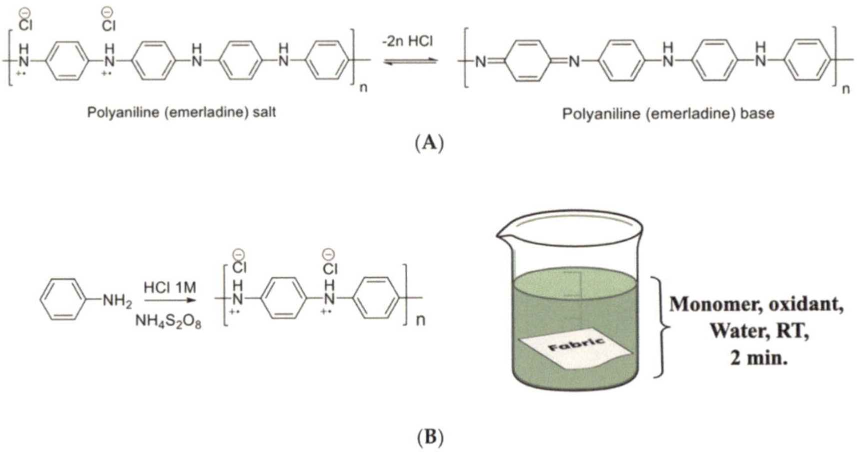

Figure 2.
Reversible transformation between polyaniline emeraldine salt (ES) and polyaniline emeraldine base (EB) (A), polymerization reaction of aniline, and schematic representation of fabric coating via in situ polymerization (B) [42].

2. Fabrication methods

Textiles are frequently coated with ICPs, such as a thin layer of PANI, to achieve electro-conductive and other functional properties. The common coating techniques are in situ polymerization, screen or inkjet printing, spray, and dip-coating [34]. The properties of the PANI-coated fabric depend on the aniline, oxidizing reagent, and dopant type and concentrations, the polymerization conditions, the pretreatment of the original fabric, and the type and style of the textile substrate.

2.1 In situ polymerization

Polyaniline can be polymerized directly on the textile surface via in situ polymerization of aniline monomer using an oxidizing reagent such as ammonium persulfate (APS) in an aqueous solution under acidic pH. During the polymerization process, the fabric is first immersed into an acidic aqueous solution containing aniline to ensure its absorption on the fabric surface for a specific time, followed by the introduction of APS triggering the polymerization reaction (**Figure 2B**). After the reaction is completed, the fabric is washed with water and organic solvents to remove the unreacted materials and byproducts, as shown in **Figure 3** [42]. Depending on the type of fabric and the presence of reactive functional groups on the fabric surface, the growing polymer chains may be attached to the fabric via physical or chemical bonding [55]. The in situ method has been successfully applied on different textile substrates, such as cotton [56], poly(ethylene terephthalate) (PET) [57], nylon [58], polypropylene (PP) [59], silk [60], and more [61–63]. Over the years, it has become one of the most popular methods for developing multifunctional electro-conductive textiles [61]. For example, Sheibani et al. carried out in situ polymerization of PANI on cotton fabrics in the presence of hydrochloric acid (HCl) as a dopant and APS as an oxidant to achieve the electrical conductivity. The electrical resistivity of PANI-coated cotton fabrics was found to be strongly dependent on the monomer concentration and the molar ratio between aniline and oxidant. When using 0.019 mol of aniline and aniline-to-dopant molar ratio of 1:7, the electrical

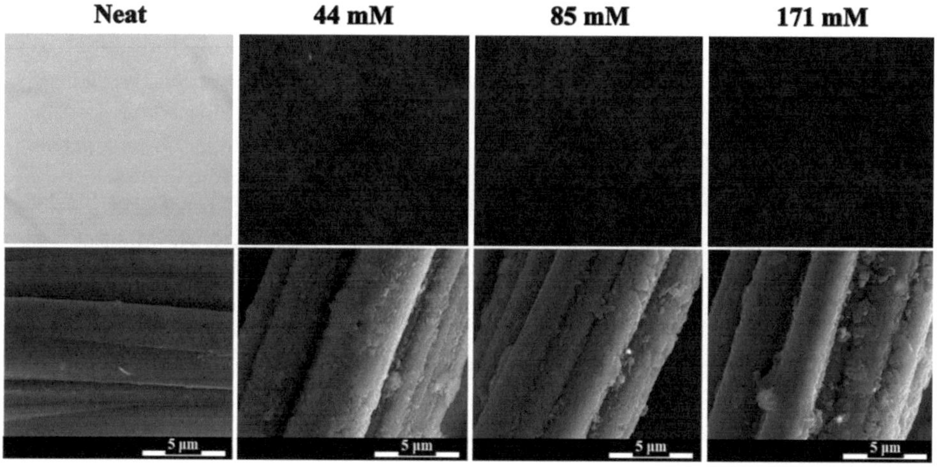

Figure 3.
Polyaniline (PANI)-coated poly(ethylene terephthalate) PET:viscose fabrics at different PANI concentrations and scanning electron microscopy (SEM) images of the fabric at 20,000× [42].

resistivity of the composite fabrics reached the level of 23×10^6 Ω/square, which was six orders of magnitude lower in comparison with the original cotton fabric [56].

Wang et al. reported the coating of woven polyester fabric by in situ polymerization using p-toluenesulfonic acid (p-TSA) as a dopant. The coated fabrics showed good polarizing and absorbing-attenuation abilities for electromagnetic shielding applications when the concentrations of aniline, oxidant, and p-TSA were 0.4 mol/l, 0.4–0.5 mol/l, and 0.2 mol/l, respectively [64].

Another option is to utilize a two-step in situ polymerization. The first stage involves wetting of the fabric in an aniline solution for several minutes or hours to ensure efficient monomer absorption on the surface of the fabric. The second stage involves either immersion or spraying of the fabric with an aqueous/acidic solution containing oxidizing reagent. Kumar et al. and Lau et al. applied the two-step fabrication method to obtain a PANI layer on woven polypropylene and non-woven fabrics with ammonia gas-sensing properties [65, 66]. Most of the studies on the fabrication of PANI-based fabrics via in situ polymerization result in the formation of physical bonding between the polar chemical groups on the surface of the fabrics and the PANI chains. Covalent grafting of PANI onto the surface of fabrics is more challenging from a synthetic point of view; however, it may offer more accurate control of the surface properties and provide coating durability. Wu et al. reported a multistep process for covalent attachment of PANI to the surface of cotton fabrics. The fabrication method included the attachment of the polymeric graft chains containing epoxy functional end-groups to the hydroxyl groups of the cellulose, which were subsequently chemically linked to the 4-amino phenethylamine molecules via an amino-epoxy ring opening reaction. The modified fabric was then subjected to in situ polymerization, creating a PANI graft layer on the surface of the cotton fabric [67].

2.2 Printing techniques

2.2.1 Screen printing

Screen printing is a well-known, versatile, low-cost printing method based on high-viscosity inks forced and squeezed through the screen in a predetermined pattern mesh. This printing technique transfers ink through the mesh onto the fabric in the desired pattern [68–70]. Consequently, screen printing is widely used to fabricate smart textiles, realizing that depositing a pattern of functional ink onto fabric creates flexible and functional coatings. The inks and the substrates have a crucial role in screen printing technology, and their properties will determine the performance of the printed fabric [71]. There are various commercial inks commonly used based on plasticized polyvinyl chloride microparticles and Speedball®, which is based on poly(methyl methacrylate) particles. Screen printing has significant potential for manufacturing wearable electronics. It is one of the most cost-effective and efficient methods for producing electro-conductive patterned coatings on different fabrics. PANI-based inks for screen printing textiles have been extensively investigated [72]. For instance, Wang et al. developed flexible circuit patterns on woven polyester fabric. Conductive pastes containing G/PANI with different concentrations of aniline between 0 and 25 w/v% and viscosity range of 3000–25,000 Pa s were successfully prepared. The highest electrical resistivity of 0.6×10^3 Ω/square was achieved when the aniline concentration was 25 w/v%. The authors found that the electrical conductivity depends on the printed width line; as the width increases, the conductivity also increases; however, it was independent of the printing time [73]. In addition, Yu et al. prepared and printed a conductive nano-silver/PANI paste on cotton fabric to form conductive circuits for flexible electronic devices. PANI was found to enhance adhesion and prevent the self-assembly of the silver nanoparticles [74]. Ma et al. prepared thermo-conductive cotton fabric by incorporating eigenstate PANI with commercial ionic liquid (1-decyl-3-methylimidazolium bromide ([DMIm]Br)) via screen printing. Although the cotton surface was rough, the inks wrapped the cotton fibers uniformly. When the coated fabric was exposed to near-infrared (NIR) irradiation at 350 mW, the electrical current was increased by 150% [75].

Furthermore, the Gray group reported screen printing of a pH electrode on polyester fabric using commercial ink combined with PANI powder and dodecylbenzene sulfonic acid as a dopant [76]. In follow-up research, the authors fabricated the first screen-printed PANI composite sensor on a textile fabric that can be applied with textile-based microfluidic channels. The schematic illustration of screen printing is shown in **Figure 4A**, and a screen-printed PANI composite electrode is shown in **Figure 4B** [77].

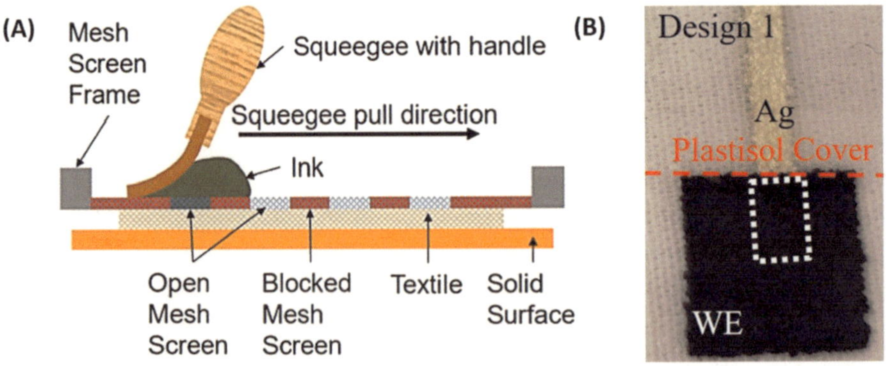

Figure 4.
(A) Schematic representation of typical textile screen printing; (B) polyaniline (PANI) composite screen printing on polyester fabric [77].

2.2.2 Inkjet printing

The inkjet printing method is attractive and versatile for controlled depositions of functional materials suitable for various surfaces. In inkjet printing, the structures are built up droplet by droplet on the surface of the substrate. Inkjet printing can be advantageous due to its low cost, esthetic finish, excellent resolution, and minimal material consumption. In comparison to screen printing, no mask is required, and there is flexibility in changing the printed pattern design depending on the final application. One of the requirements of this technique is to use low-viscosity fluids having viscosity values under 20 centiPoise (cP) and surface tension between 28 and 350 Nm^{-1} [78, 79]. Until today, this processing method has not been widely explored for PANI-coated fabrics, with only a limited number of studies reported in recent years. Stempien et al. proposed a facile method based on sequential line-by- line reactive inkjetting of aniline monomer and APS on polyacrylonitrile (PAN), cotton, PET, cotton/PET, wool, and cotton/wool fabrics for electromagnetic interference (EMI) shielding applications. The electro-conductive lines were printed on the surfaces of the fabrics by a selected-controlled pattern having the first nozzle print the aqueous solution of aniline hydrochloride and immediately after, the second nozzle prints the aqueous solution of ammonium persulfate or vice versa, as shown in **Figure 5** [80, 81].

2.3 Spray-coating

In the spray-coating technique, the fabric can be sprayed with aniline solution directly or polymerized by immersion in aniline solution, followed by spraying an APS solution. By using a screen mesh, one can control the pattern of the conductive

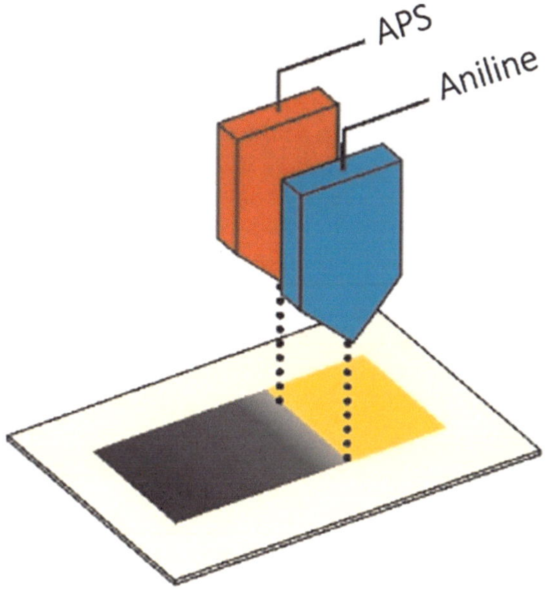

Figure 5.
Schematic representation of inkjet-printed textile of the polyaniline (PANI) layer [80].

particles or droplets on the textile substrate. Similar to screen-printing and inkjet-printing techniques, spray-coating has yet to be extensively investigated in fabricating PANI-based conductive textiles [79]. Qin et al. developed a multifunctional protective spray-coating based on a layer-by-layer assembly of PANI and GO on cotton fabrics. The GO particles were efficiently dispersed in the PANI nanowire arrays, affording nanoscale coating with controllable thickness and excellent antistatic, self-extinguishing, and antimicrobial properties [82]. Yu et al. also fabricated electrically conductive PANI/two-dimensional (2D) metal carbide (MXene)/cotton fabrics by a spray-coating method for acid/alkali-responsive and tunable electromagnetic interference (EMI) shielding applications, as shown in **Figure 6** [51].

2.4 Dip-coating

Dip-coating is a common approach for conductive textile materials due to its low cost, simplicity, and scale-up ability. This technique requires no specialized equipment, making it widely used [21]. The dip-coating method demonstrates a thin electro-conductive layer created on the surface of the fabric by immersion into an aqueous acidic solution containing PANI [83]. One of the drawbacks of this method is the formation of non-uniform coatings limiting the control of the coating thickness [84]. Various processing parameters, such as temperature, solution concentration, dip-coating time, and withdrawal speed, strongly impact the properties of the coating [83, 85]. There are several reports on the fabrication of PANI-coated fabrics via the dip-coating technique. Mahat et al. used dip-coating to fabricate a conductive polyaniline cotton fabric with phosphoric acid and p-TSA as dopants. The results revealed an increased thickness layer of PANI and improved electro-conductivity with increased dopant concentration [47]. The authors later described the fabrication of PANI-coated polyester fabric via dip-coatings that exhibited good antibacterial properties (**Figure** 7) [52]. Another study presented the preparation of thermo-electric polyester/linen woven by dip-coating into acetone suspension containing PANI/graphene nanosheets (GNSs). The preparation method included ultrasonication to establish efficient dispersion of GNS in acidic solution prior to the introduction of the aniline monomer and APS [86]. Wang et al. utilized dip-coating for the fabrication of a flexible gas sensor based on PANI and graphene nanoplatelets on PP fabric. In this case, the fabric was first dipped into graphene suspension, which was prepared via ultrasonication and addition of sodium dodecylbenzenesulfonate to enhance the

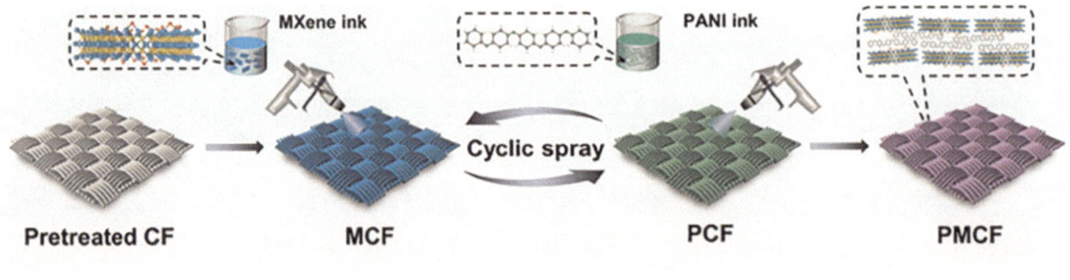

Figure 6.
Schematic representation of fabrication of electrically conductive PANI/MXene/cotton fabrics by spray coating [51]. Copyright © 2022, American Chemical Society.

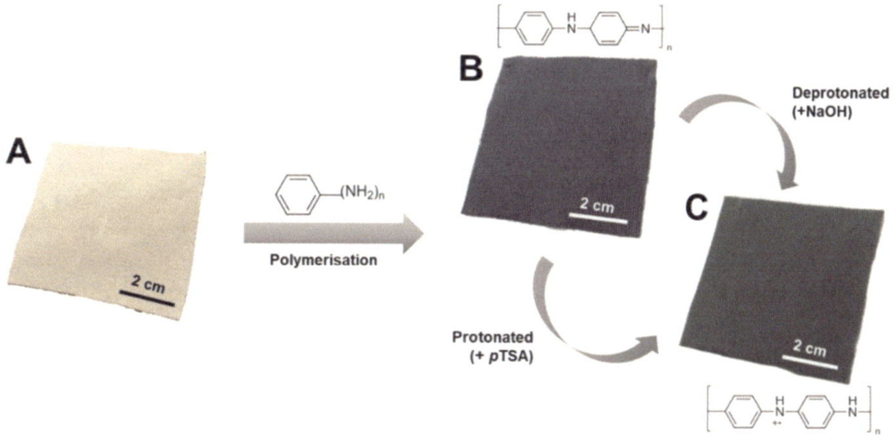

Figure 7.
Polyaniline (PANI) dip-coating process on polyester fabric [52].

graphene dispersion, followed by dipping into a PANI solution. The sensor fabric demonstrated a detection limit of 100 parts per billion (ppb) for ammonia gas [61]. Alamer et al. prepared a dip-coated PANI/carbon black (CB) composite on cotton fabric. Increasing PANI/CB concentration showed deeper coating penetration between the fibers [87].

3. Properties and applications

3.1 Electrical conductivity

Electrical conductivity is an important characteristic of PANI-coated textiles [88]. The electronic properties of PANI are associated with the emeraldine oxidation state and can be altered by reversible oxidation/reduction and protonation/deprotonation processes. The molecular weight and color of the PANI depend on the chemical composition of the reactants, concentration, and reaction conditions used for its synthesis. The parameters affecting the electrical conductivity of PANI-coated fabrics include the amount of polymer deposition, density, structure, and nature, which is evident due to a high number of picks per inch, resulting in lower surface resistivity [89, 90]. For example, Madhusoothanan et al. investigated the electrical resistivity of PANI/PET fabrics prepared via in situ polymerization with different weaving patterns of twill, satin, and plain, and 50–80 picks per inch. Their results revealed that among 80 picks per inch fabrics, the twill fabric showed lower surface resistivity of 1300 Ω/square compared to satin and plain fabrics that exhibited resistivity values of 1340 Ω/square and 1644 Ω/square, respectively. In addition, a reduction in surface resistivity was observed with increased fabric density for all weaving types [91]. Cellulose-based fabrics show efficient deposition of PANI due to the hydrogen bonds created between the aniline monomer and the surface of the yarns. This is reflected by a significant reduction in the electrical resistivity of the coated cotton-based fabrics compared to the untreated ones. For example, Hong et al. reported an electrical resistivity of 0.275×10^3 Ω cm obtained for knitted cotton fabric (255 grams per square meter

(GSM)), which was eight orders of magnitude lower than the pristine fabric [92]. The concentration of aniline used for the polymerization also affects the electrical resistivity of the coated fabrics, which is decreased upon increasing the monomer concentration [42, 56, 93]. The type of dopant used for the polymerization of PANI may also affect the electrical properties of the fabrics. For example, using hydrochloric acid, hydrofluoric acid (HF), and hydrobromic acid (HBr) for aniline polymerization on cotton, wool, and polyester woven fabrics, the highest conductivity was observed for the bromine-doped PANI-coated fabrics [23]. A further enhancement in electrical conductivity of the PANI-coated fabrics may be achieved by combining PANI with carbon-based additives such as carbon black, carbon nanotubes, graphene, and graphene oxide, or conductive polymers such as polypyrrole and metal oxides such as MnO_2 [40, 63, 87, 94–100]. To improve the interaction between PANI and the surface of the fabric to provide a good distribution of the conducting coating compositions, various compounds, such as polystyrene sulfonate (PSS) [101, 102], chitosan (CTS) [103], poly(2-methoxyaniline-5-sulfonic acid) (PMAS) [104], nanodiamond [20], polyetherimide [61], and 4-amino phenethylamine-glycidyl methacrylate [67], were also introduced. These materials contain polar chemical groups that establish efficient secondary interactions, such as hydrogen bonding with the aniline monomer, and effectively bridge between PANI molecules to form a continuous electro-conducting network. **Table 1** summarizes the electrical properties of PANI-coated textiles with various coating compositions and their applications.

3.2 Polyaniline-based textile sensors

Due to the unique chemical structure of PANI, it can act as a stimuli-responsive material, responding to changes in the environment by changing its color and electrical conductivity. Consequently, PANI-coated fabrics have been extensively studied as sensors detecting different chemicals, pressure, or mechanical changes in the environment [6, 26, 34, 107, 108].

Textile type and coating composition	Fabrication method	Electrical conductivity/ resistivity	Application	Ref.
PANI/cotton woven (143 GSM)	In situ polymerization, HCl, ammonium persulfate (APS)	71×10^6 Ω/square	Electro-conductive textile	[56]
PANI/silk crepe, PANI/ reduced graphene oxide (RGO)/silk crepe	In situ polymerization, HCl, APS	1.92×10^3 Ω/cm, 0.19×10^3 Ω/cm	Wearable electronic textile	[93]
PANI/wool, PANI/PET, PANI/nylon 6, PANI/acrylics, PANI/cotton	In situ polymerization, HCl, potassium iodide (KI), APS, potassium bichromate	10^4–10^6 Ω/square	Electro-conductive textile	[63]
PANI/cotton knitted (255 GSM)	In situ polymerization, HCl, APS	0.275×10^3 Ω·cm	Strain sensor	[92]
PANI/nylon 6 woven	In situ polymerization, HCl, APS	0.6×10^{-1} S/cm	Highly electro-conductive clothing	[105]

Textile type and coating composition	Fabrication method	Electrical conductivity/ resistivity	Application	Ref.
PANI/PAN (220 GSM), PANI/cotton (120 GSM), PANI/wool (150 GSM), PANI/cotton/PET (50/50) (250 GSM), PANI/cotton/ wool (25/75) (180 GSM), All woven fabrics	Reactive inkjet printing, water, APS	233 Ω/square, 702 Ω/ square, 1570 Ω/ square, 1550 Ω/ square,438 Ω/square	Electro-conductive textile in electromagnetic interference shielding	[81]
PANI/PET, PANI/PET/ viscose (70/30), PANI/bamboo, PANI/ bamboo/cotton (70/30), All woven fabrics	In situ polymerization, HCl, APS	29 Ω·m, 163.2 Ω·m, 12.4 Ω·m, 8.9 Ω·m, 6.2 Ω·m	Electro-conductive textiles	[90]
PANI/viscose (170 GSM), PANI/PET (150 GSM), PANI/PET:viscose (50:50) (370 GSM), All non-woven fabrics	In situ polymerization, HCl, APS	5.0×10^7 Ω/square, 10×10^6 Ω/square, 1.6×10^5 Ω/square	Electro-conductive and antibacterial textile	[42]
PANI/cotton (192 GSM) woven	In situ polymerization, HCl, HF, and HBr, APS	3281 μS/cm (HCl), 2071 μS/cm (HF) and 3314 μS/cm (HBr)	Electro-conductive textile	[23]
PANI/ 4-aminophenethylamine-glycidyl methacrylate/cotton	Covalently grafting by multistep treatment process, APS	1×10^7 Ω/square	Multifunctional electro-conductive textiles	[67]
PANI/Ag/cotton (120 GSM)	In situ polymerization, aniline hydrochloride, APS	19×10^6 Ω/square	Electrostimulation or monitoring of wound dressing	[106]
PANI/MnO_2/cotton	In situ polymerization, HCl, APS	1.43×10^{-1} $\Omega^{-1}\cdot m^{-1}$	Energy storage device	[99]
PANI/CB/cotton	Dropcasting, dimethylsulfoxide (DMSO)	0.494×10^3 Ω/ square	Electro-conductive textile	[87]
PANI/GO/cotton woven	In situ polymerization, HCl, APS	48.35 Ω·cm	Electro-conductive and UV-blocking textile	[98]
PPy/PANI/cotton woven (120 GSM)	In situ polymerization, aniline hydrochloride, APS	210 Ω/square	Monitoring of carnivorous plant response	[97]
PANI/ ramie woven (70 GSM), PANI/PEI/ramie	In situ polymerization, HCl, APS	16×10^6 Ω/square, 0.35×10^6 Ω/square	Electro-conductive multifunctional textiles	[61]
PANI/carbon nanotubes (CNTs)/G/PET woven	In situ polymerization, H_2SO_4, APS	114 Ω/square	Supercapacitor electrode textile	[100]
PANI/chitosan/ wool (80 GSM)	Two-step polymerizations, HCl, APS	11.32 S/cm	Electro-conductive and antibacterial textile	[103]
PANI/PMAS/wool-nylon-lycra	In situ polymerization, HCl, APS	342×10^3 Ω/square	Strain sensor	[104]

Textile type and coating composition	Fabrication method	Electrical conductivity/ resistivity	Application	Ref.
PANI/boron-doped nanocrystalline diamond (BDND)/wool knitted (180 GSM)	In situ polymerization, HCl, APS	6.4×10^3 Ω/square	Wearable strain sensor	[20]
PANI/polystyrene sulfonate (PSS)/polyamide (112 GSM), PANI/PSS/wool (184 GSM)	In situ polymerization, water, APS	1.4×10^{-4} S/cm, 1.1×10^{-4} S/cm	Sensor-based textile	[101]
PANI/PSS/nylon woven	Layer-by-layer (LBL) assembly, in situ polymerization, p-toluenesulfonic acid (p-TSA), APS	1.7×10^{-5} S/cm	Electro-conductive textile	[102]
PANI/polyester (plain, twill, satin)	In situ polymerization, HCl, APS	1644 Ω/square, 1300 Ω/square, 1340 Ω/square	Electro-conductive textile	[91]

Table 1.
Electro-conducting properties and applications of polyaniline (PANI)-coated textiles.

3.2.1 Gas sensors

Polyaniline-coated fabrics are promising candidates for various gas-sensing applications due to the reversible acid/base reactions with gas molecules accompanied by the change in color and electrical conductivity [109–111]. **Figure 8** shows a chemical reaction of a quaternary ammonium moiety of PANI-ES forming PANI-EB [110–112]. Extensive research has been focused on the development of PANI-coated fabrics as sensors for the detection and monitoring of ammonia, methanol, ethanol, acetone, hydrogen sulfide (H_2S), nitrogen dioxide (NO_2), and other gases [113–115]. Essential parameters for PANI-based gas sensors are the detection limit that defines the lowest concentration of the gas that can be reliably detected, and the "selectivity" refers to the

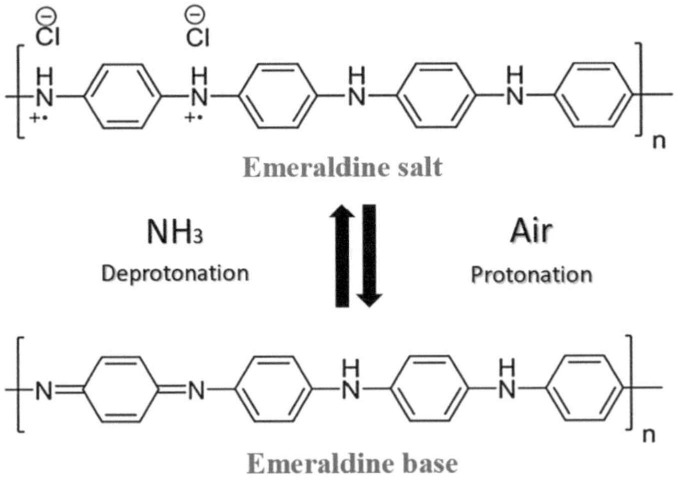

Figure 8.
Acid-base reaction between polyaniline (PANI) and ammonia (NH_3).

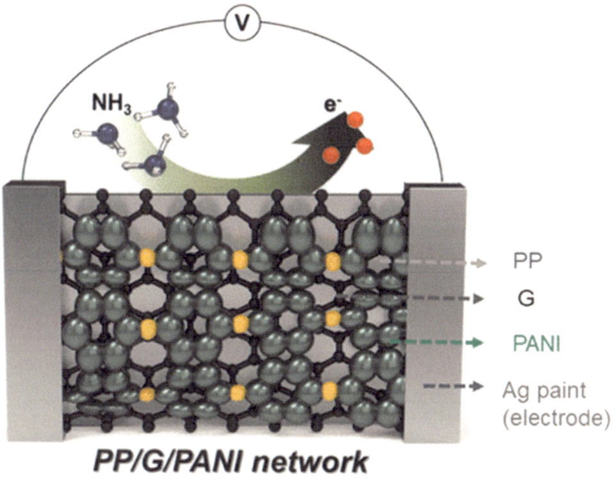

Figure 9.
Sensing mechanism of the PP/G/PANI hybrid sensor [59]. Copyright © 2022, American Chemical Society.

ability to distinguish between various gases [116]. For instance, Jia et al. fabricated PANI on cotton fabric for detecting ammonia gas in the range of 1–20 parts per million (ppm), while the detection of H_2S in the range between 1 and 1000 ppm was studied by de Souza et al. [117, 118]. The incorporation of additives, such as graphene, graphene oxide [50], and MXene [51], into PANI coating was shown to significantly improve the diffusion and penetration depth for the analyte gas molecules [110, 111, 119–124]. Kim et al. fabricated a hybrid sensor based on PANI/G/PP, allowing the detection of ammonia gas from 100 ppb to 75 ppm with a detection limit of 1 ppm. When the sensor was exposed to 50 ppm of ammonia, the response and recovery times were 114 s and 23 s, respectively. The sensor also sensed H_2S gas in the range of 10–50 ppm, with a response time of 138 s and a recovery time of 22 s for 50 ppm (**Figure 9**) [59]. Another gas sensor was fabricated by spraying PANI and multiwall carbon nanotubes (MWCNTs) on the PP fabric. The sensor was characterized for detecting ammonia in the 20–100 ppm range and demonstrated a detection limit of 200 ppb, as presented in **Figure 10a**. Furthermore, compared to nine other organic protic and aprotic solvents, the sensor

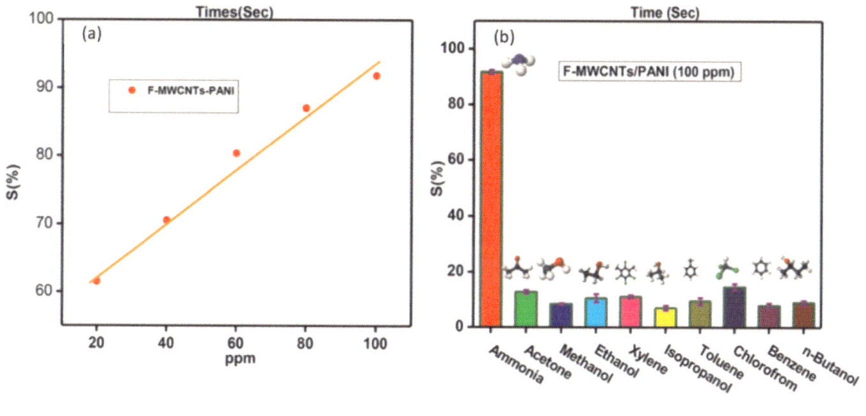

Figure 10.
Sensor response: (a) sensor response plot of F-MWCNTs/PANI toward the exposure of ammonia; (b) selectivity of the F-MWCNTs/PANI [65]. Copyright © 2018, American Chemical Society.

Sensor fabric	Fabrication method	Gas type	Detection range [ppm]	Response value [ppm]	Response time [sec]	Recovery time [sec]	Ref.
PANI/cotton	In situ polymerization	NH_3	25–100	—	—	—	[113]
PANI/nylon 6	In situ polymerization	NH_3	1000	—	—	—	[125]
PANI/cotton	In situ polymerization	NH_3	100	100	1.9	30	[117]
PANI/PP	Inkjet printing	NH_3	15–100	—	—	—	[114]
PANI/PET	In situ polymerization	NH_3	29	—	—	—	[115]
		NO_2	69				
PANI/cotton	In situ polymerization	H_2S	1–1000	—	—	—	[118]
PANI/G/PP	In situ polymerization	NH_3	0.01–75	50	114	23	[59]
		H_2S	10–50	50	138	22	
PANI/ MWCNTs/PP	In situ polymerization	NH_3	20–100	100	9	30	[65]
V_2O_5/PANI/ GO/PET	In situ polymerization	NH_3	0.5–100	20	70	520	[50]
				1	78	259	
PANI/MXene/ cotton	Spray-coating	NH_3	10–200	50	308	—	[51]
PANI/WO_3/ cotton	In situ polymerization	NH_3	3–150	100	122	165	[119]
PANI/clean wiper	In situ polymerization	NH_3	10–320	20	50	500	[120]

Table 2.
Textile gas sensors based on polyaniline (PANI) coatings.

showed more significant selectivity for the ammonia gas, as shown in **Figure 10b** [65]. **Table 2** summarizes the composition, fabrication methods, and gas-sensing characteristics of the main reported textiles-based gas sensors.

3.2.2 pH sensors

Textiles have become an attractive platform for fabricating pH sensors in the past decade due to their breathability, flexibility, and bending ability. Consequently, PANI-based textile sensors are proposed for monitoring skin pH for different skin conditions and diseases. Similar to the gas detection mechanism, depending on the pH values of the environment, the nitrogen atoms of the PANI backbone undergo protonation/deprotonation reactions between the imine groups and the quaternary ammonium ions. These processes allow rapid and reversible transformation between PANI-ES and PANI-EB in the pH range of 2–12 that are obtained by the changes of color, electrical and optical properties (**Figures 8** and **11**) [126–129]. For example, Gray et al. fabricated a pH sensor based on screen printing of PANI-incorporated commercial ink on polyester fabric. The sensor properties were evaluated in the 3–10 pH range and showed a sensitivity of −28.2 mV/pH [76]. Other studies showed that

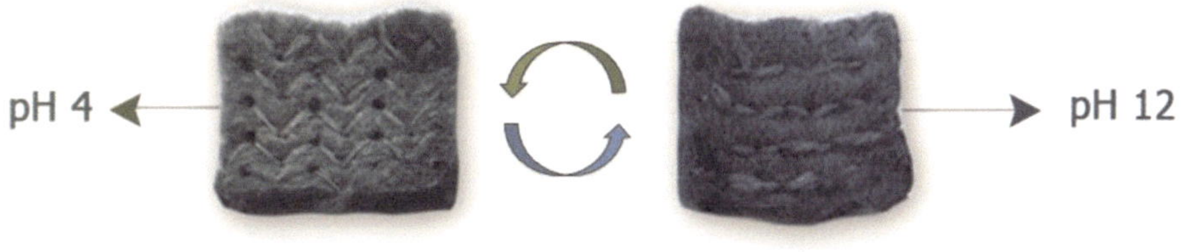

Figure 11.
Polyaniline (PANI)-coated non-woven cotton fabric functional as an optical pH sensor.

PANI emeraldine salt had a better sensitivity of −42.6 mV/pH than PANI emeraldine base, which had a sensitivity of −27.9 mV/pH [77]. Koo et al. fabricated a skin pH sensor using a paste composed of carbon nanotubes (CNTs), agarose, and PANI on cotton, regular polyester, waterproof polyester, and thermoplastic polyurethane as an overlayer to improve stability. The sensor displayed a linear dependence on pH ranging from 5 to 9 with a 45.9 mV/pH sensitivity and maintained its sensing ability after repeated bending cycles [130]. Another wearable pH sensor was prepared from PANI combined with chopped carbon fibers and showed a Nernstian response of 58.0 mV/pH in the presence of pH values in the 4–8 range [131].

3.2.3 Mechanical sensors

Wearable sensors can interact with the human body/skin and monitor the movements of different body parts as well as physiological signals such as muscle vibrations and pulse [132–134]. Textile-based PANI sensors are divided into strain (triboelectric) and pressure (piezoelectric) sensors, responding to the corresponding stimuli by changing their electrical conductivity [132, 135]. Mechanical sensors can monitor stress, torque, pressure, or force applied to a textile [136–138]. Textile-based strain and pressure sensors show a good correlation between the applied mechanical deformation and the change in the electrical resistivity [139–141]. When strain is applied to the fabric, it causes the straightening of the yarns, increasing the density, improving the solid-state packing of the PANI molecular chains, and increasing the electrical conductivity of the coated fabric due to more efficient charge transport. Upon structural relaxation of the fabric, the free volume between PANI molecules increases, decreasing the electrical conductivity, as shown in **Figure 12** [143, 144]. As a result, testing electrical conductivity changes as a function of the applied strain shows that the PANI-coated fabric sensors exhibit high strain linearity. The performance of the strain sensor depends on the weaving structure and intrinsic elasticity of the fabrics as well as the ability to form an efficient coating with strong interactions between the PANI chains and the fabric, enabling it to withstand tensile deformation. For example, Hong et al. developed a strain sensor based on PANI/cotton knitted fabric showing good repeatability for strain values up to 10% with excellent stretch recovery of 1000 cycles. When the strain was increased to 20%, the sensing repeatability was affected by the damaged PANI layer [92]. Kannaian et al. introduced a strain sensor based on PANI/nylon-lycra (92:8) fabric that exhibited changes in electrical conductivity up to 50% strain due to typical human body movement and bending angles up to 90°. Above these angles, they have yet to get a significant response. The gauge factor represents the sensitivity of a strain sensor, which was 0.92 [145]. Another PANI-based wearable strain sensor was fabricated using a 4-way stretch nylon-spandex (80:20) tricot fabric encapsulated with waterborne polyurethane coating to maintain the PANI onto the fabric. The sensor can

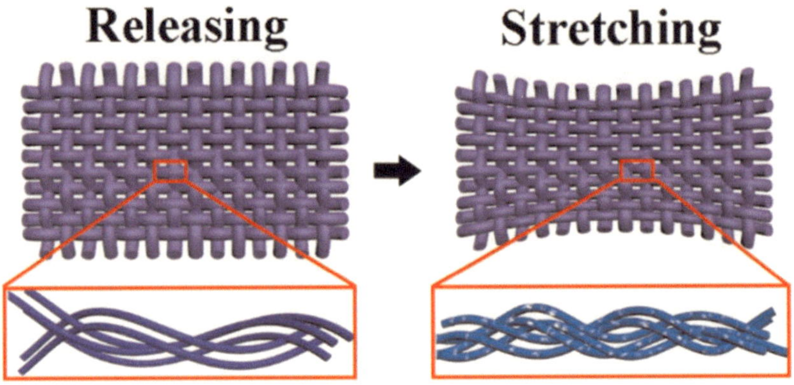

Figure 12.
Schematic illustration of the structural change in the conductive polymers (CPs)-coated polyester/spandex textile under a stretching force [142].

monitor different knee states, such as walking and running, hand movements, and breathing rates in a strain range between 2.5 and 30%, with a gauge factor of 1–2.5 while detecting loadings up to 50 g. In addition, the sensor showed the ability to endure repeating 10% strain for 1000 cycles with a relative resistance change (ΔR/R) between 2.5 and 23% [146]. Another study described a strain sensor fabricated on lycra fabric using PANI/graphene nanoplatelets/silicon rubber successfully monitored up to 40% strain with a change of ΔR/R between 0 and 30%, and a gauge factor of 67.3. In addition, the sensor could monitor the bending angle of a finger up to 120° [147]. Composite coatings containing PANI and carbon nanotubes or titanium dioxide were fabricated on milk protein-based and knitted polyester fabrics operated at an even larger strain scale of 50–60% [62, 148]. **Table 3** summarizes the composition, fabrication methods, and strain characteristics of the main reported textiles-based strain sensors.

Sensor fabric	Fabrication method	Operating regime/sensing range [%]	Resistance	Cycling stability [cycles]	Monitoring ability	Ref.
PANI/cotton knitted	In situ polymerization	0–20	—	1000	Elbow, knee, finger, and, laryngeal	[92]
PANI/nylon-lycra (92:8)	In situ polymerization	0–60	12×10^3– 11×10^3 Ω	25	Elbow and knee	[134]
PANI/nylon-lycra (92:8)	In situ polymerization	0–50	22×10^3– 15×10^3 Ω	5	Elbow	[145]
PANI/ nylon-spandex (80:20)	In situ polymerization	0–25	5.1×10^3– 3.8×10^3 Ω	1000	Knee and hand movement	[146]
PANI/GNPs/ SR/lycra	Spin-coating	0–40	2.23×10^3– 5×10^2 Ω	40	Finger	[147]
PANI/TiO_2/ polyester	In situ polymerization	0–60	—	100	—	[148]
PANI/CNTs/ protein	In situ polymerization	0.1–50	—	—	Laryngeal, finger, elbow, and knee	[62]

Table 3.
Textile strain sensors based on polyaniline (PANI) coatings.

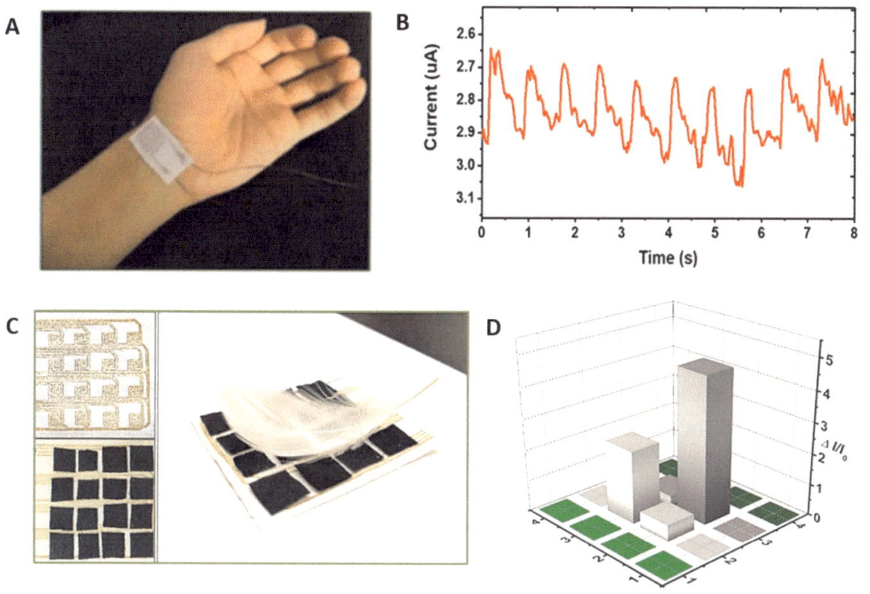

Figure 13.
(A) Real-time detection of physiological signals of a wrist by using the pressure sensor-based polyaniline (PANI). (B) Detection of a wrist pulse current. (C) A screen-printed sensor-based PANI and feather sensing. (D) By three-dimensional (3D) bar graph that displays the current change dependent on the pressure area [149].

Li et al. reported a pressure sensor made of PANI-coated non-woven cotton fabrics and screen-printed electrodes with an excellent sensitivity of 46.48 kPa^{-1} in a pressure range of up to 4.5 kPa. The sensor showed a low detection limit of 0.46 Pa, fast response and relaxation time of 7 and 16 ms, respectively, and high stability for more than 250 cycles. They tested the detection of physiological signals, such as wrist movement shown in **Figure 13A**, and the plot in **Figure 13B** demonstrates the change in current between 2.75 and 3.05 μA. **Figure 13C** presents the fabrication of silver paste electrodes by screen printing on neat and PANI-coated cotton placed together using a feather weight to create a ΔI/I change, as shown by a 3D graph in **Figure 13D** [149].

Additionally, Zeng et al. designed a PANI-coated cotton sensor for monitoring finger pressing and fingertip motion with normalized current change between 0 and 4 and 0–6, respectively. The sensor senses between 300 Pa and 30 kPa, with a response time of 0.4 s and a relaxation time of 0.2 s [120]. Yu et al. prepared a PANI/Ag/PET pressure sensor with high sensitivity and good linearity for monitoring walking, running, squatting, and jumping. The sensor showed an immediate response at pressures between 2.5 and 6.5 kPa with a sharp increase in the relative change in the resistance (ΔR/R) due to an increase in the effective contact area [150]. Yu et al. prepared a PANI/nano-silver/cotton mechanical sensor. The sensor showed a sensitivity of 0.04–0.10 kPa^{-1} with a sensing range of 0–20 kPa, and a quick response and recovery time of 0.4 s. The sensor demonstrated 500 loading and unloading cycles with a resistance of 3.5×10^6 Ω and 1.1×10^6 Ω, respectively. Preparing a fabric with a layer-by-layer structure of PANI and nano-silver increases the electro-conductivity and contributes to excellent properties. This sensor can effectively monitor the movements of vocal cord vibrations, arm, elbow, and foot [151]. **Table 4** summarizes the composition, fabrication methods, and pressure characteristics of the main reported textiles-based pressure sensors.

3.3 Electromagnetic interference shielding

Electromagnetic interference shielding can protect electronic equipment and humans against external electromagnetic irradiation. EMI shielding is based on three components: reflection, absorption, and multiple reflections inside the shielding material at different frequency ranges [152]. Textiles are useful for protecting materials due to their light weight, flexibility, and drapability. Textile-based EMI shielding is strongly driven by using new materials, such as ICPs, to fabricate protective coatings [153, 154]. In this emerging field, PANI-based textiles have gained considerable research interest due to their tunable electronic properties combined with the ease of processability and the ability to control the thickness and the amount of shielding coating. PANI can be used solely or in combination with other conjugated polymers, metal particles, or carbon-based additives. EMI shielding properties are also affected by other factors, such as the fabric structure, porosity, and thickness.

For example, Ozdemir et al. reported an average electromagnetic shielding efficiency of 3.8 dB and an average absorption value of 48% for PANI-coated cotton fabrics [155]. Also, Sun produced a PANI-coated cotton fabric with a low surface resistivity of 0.5 kΩ and electromagnetic shielding effectiveness of up to 60%. They demonstrated the correlation between the oxidant concentration and the shielding effectiveness, resulting in a decreased electromagnetic shielding effectiveness when increasing the APS concentration [156]. In addition, Jia et al. prepared PANI-coated cotton composite strips with ammonia gas-sensing and EMI shielding properties. The EMI shielding effectiveness varied depending on the frequency and increased to about −9 dB in the frequency range of 4.5 GHz and 9 GHz. [117]. Muthukumar et al. investigated the correlation between the fabric type and the EMI shielding effectiveness. They displayed cotton, polyester, and nylon PANI-coated fabrics in the frequency range of 8–12 GHz with EMI values of −1.62, −2.78, and −1.5 dB, respectively [157]. In another study, Kannaian et al. developed a conductive PANI-coated nylon-lycra fabric with an EMI shielding efficiency of 26 dB in the frequency range of 8–12 GHz [158]. Dhawan et al. fabricated PANI coating on polyester fabric with various PANI layer thicknesses, showing a strong correlation between the coating thickness layer and the shielding effectiveness. The highest shielding effectiveness of 21 dB was obtained for a three-layered PANI-coated polyester fabric, while a PANI-coated silica fabric exhibited a shielding effectiveness of 35 dB at 101 GHz [159]. The EMI shielding can be enhanced by fabricating PANI composite coatings with other conductive materials. For example, Yu et al. used a spray-coating layer-by-layer assembly to prepare a conductive PANI/MXene/cotton fabric with an EMI shielding effectiveness of 54 dB, compared to 45 dB, which was obtained for an MXene/cotton fabric shown in **Figure 14**. The researchers found a direct relationship between the number of spray cycles and EMI shielding properties. It was shown that acid/alkali environments can trigger EMI shielding. When treated with an alkali gas, the shielding effectiveness decreased to 15 dB, while exposure to acidic gas increased the EMI shielding effectiveness to 22 dB [51]. Yu et al. reported a PANI/nickel-tungsten-phosphorus (Ni-W-P) coating on functional polyimide fabric with an electrical resistivity of 0.08 Ω/square and an outstanding EMI shielding effectiveness of 103 dB. This study highlights the relationship between high electrical conductivity and excellent shielding performance [58]. Additionally, Sebastian et al. incorporated graphite to

Sensor fabric	Fabrication method	Pressure range	$(I-I_0)/I$	$(R-R_0/R)$	Response time/ Recovery time	Compression force and frequency for cycling stability measurement	Cycling stability	Monitoring ability	Ref.
PANI/cotton	In situ polymerization	0.5–30 N	—	—	—	5 N and 5 Hz	2000	Leaning back, sitting, and hand	[133]
PANI/cotton	In situ polymerization	4–30 kPa	0.5–16	—	0.4 s/0.2 s	6 kPa and 0.5 Hz	1000	Fingers	[120]
PANI/cotton	In situ polymerization	0–30 kPa	0–800	—	7 ms/16 ms (2 kPa)	5 kPa and 0.05 Hz	250	Wrist and neck	[149]
PANI/Ag/PET	In situ polymerization	2.5–9 kPa	—	2.5–65	0.3 s/0.2 s (9 kPa)	9 kPa	1300	Walking, running, squatting, and jumping	[150]
PANI/nano-silver/cotton	Printing	0–8 N	—	0–60	0.4 s/0.4 s	—	500	Vocal cord, elbow, wrist, and foot	[151]

Table 4.
Textile pressure sensors based on polyaniline (PANI) coatings.

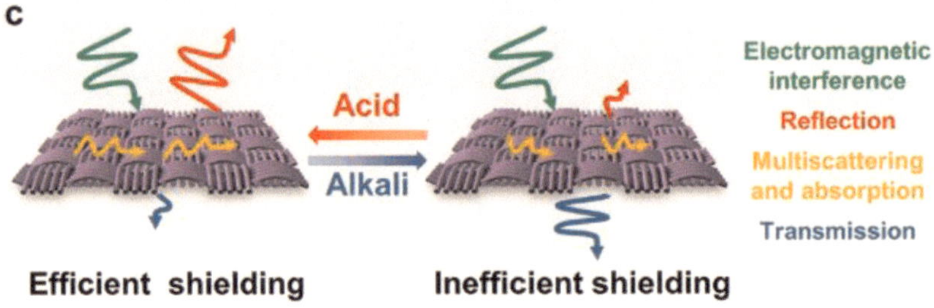

Figure 14.
Schematic draw of acid/alkali-responsive and tunable EMI shielding behavior of the PANI/MXene/cotton fabric [51]. Copyright © 2022, American Chemical Society.

enhance the shielding properties of PANI nanofibers due to their conductivity and flake-like morphology. They tested the PANI powder and the PANI/graphite powder to show the effect of graphite additive on the electrical conductivity and shielding effectiveness. The results showed increased electrical conductivity and shielding effectiveness from 18.5 S/cm to 24 S/cm and from 71 to 77 dB to 83–89 dB, respectively. Incorporating PANI nanofibers/graphite composite coating with a thickness of about 0.1 mm into cotton and nylon fabrics resulted in 11–15 dB shielding effectiveness in the frequency range of 8.2–18 GHz [160]. Another study described PANI/silver-plated PET fabrics with surface resistivity of 0.1 Ω/square, with shielding effectiveness between 60 and 90 dB in the frequency range between 30 kHz and 3 GHz, which was barely affected after ultrasonic washing for 30 min (50–90 dB) [57]. **Table 5** summarizes the composition, fabrication methods, and shielding effectiveness characteristics of the main reported textiles-based EMI shielding.

Sensor fabric	Fabrication method	Electrical conductivity/ resistivity	Frequency range	EMI shielding effectiveness	Ref.
PANI/cotton	In situ polymerization	350 Ω	6–14 GHz	3.8 dB	[155]
PANI/cotton	In situ polymerization	0.5×10^3 Ω	1–1500 MHz	60%	[156]
PANI/cotton	In situ polymerization	—	0.3–9000 MHz	0–(−9) dB	[117]
PANI/MXene/cotton	Spray-coating	0.64 Ω/square	8.2–12.4 GHz	~54 dB	[51]
PANI/cotton, PANI/ polyester, and PANI/ nylon	In situ polymerization	7×10^3 Ω/square, 5×10^3 Ω/square, 5×10^3 Ω/square	8–12 GHz	−1.62, −2.78, and − 1.5 dB	[157]
PANI/nylon-lycra (92:8)	In situ polymerization	3.5×10^3 Ω/square	8–12 GHz	26.38 dB	[158]
PANI /polyester and PANI /silica	In situ polymerization	26 Ω·cm, and 20–28 Ω ·cm	101 GHz	21.48 dB and 35.61 dB	[159]
PANI/artificial suede	In situ polymerization	9.2 S/m	8.2–12.4 GHz	25.90 dB	[161]

Sensor fabric	Fabrication method	Electrical conductivity/ resistivity	Frequency range	EMI shielding effectiveness	Ref.
PANI/nickel-tungsten-phosphorus/ polyimide	In situ polymerization	0.08 Ω/square	0.3–3000 MHz	103 dB	[58]
PANI/graphite/ cotton and PANI/ graphite/nylon	In situ polymerization	—	8.2–18 GHz	15 dB and 14 dB	[160]
PANI/silver-plated/ PET	In situ polymerization	0.1 Ω/square	0.3–3000 MHz	50–90 dB	[57]

Table 5.
Polyaniline (PANI)-based fabrics with electromagnetic interference (EMI) shielding properties.

3.4 Antibacterial textile

Over the past decade, extensive research has been focused on developing textiles that can either kill or inhibit the proliferation of bacteria, fungi, and viruses [162]. Microorganisms can easily multiply on certain types of fabrics, causing odors, hygiene problems, and a reduction in the mechanical strength of the fabric. Furthermore, the increased spreading rate of epidemics brought the need for antibacterial fabrics for their use in public places and personal care. Consequently, dozens of studies have been introduced describing the incorporation of various antibacterial agents into fabrics, including nanoparticles of silver, zinc, and magnesium oxides, essential oils, chitosan, and chitin [42, 162–167]. The exact mechanism of the antibacterial effect of these and other reagents is unclear. However, the one proposed mechanism involves electrostatic interactions between cationic moieties of the antibacterial reagent and negatively charged surface or RNA/DNA proteins of the bacterial membrane [42, 52, 168, 169]. Emeraldine salt of polyaniline is an excellent material for antibacterial applications since it may be considered a poly quaternary ammonium ion containing many effective cationic sites within a single polymer chain, as shown in **Figure 15** [42]. In addition, PANI is insoluble in water and can establish strong intermolecular interactions with the fabric providing coating durability, unlike other organic quaternary ammonium salt-based reagents and N-halamines that are water soluble and can be removed upon repeated washings.

Furthermore, a large quantity of small molecule-based antibacterial reagents will be required to achieve the effect compared to a small quantity of PANI, which can be sufficient due to its polymeric nature. The transformation of PANI-ES and PANI-EB is accompanied by a color change from green to blue, which can provide a visual indication of the antibacterial effectiveness and the possibility of restoration by re-doping. The antibacterial properties of PANI-coated cotton fabric were first reported by Bhat et al. [170]. Since then, a significant number of publications have been published describing PANI-based fabrics with antimicrobial effects against a variety of Gram-negative and Gram-positive bacteria, including *Escherichia coli* (*E. coli*), *Staphylococcus epidermidis* (*S. epidermidis*), *Staphylococcus aureus* (*S. aureus*), *Pseudomonas aeruginosa* (*P. aeruginosa*), *Enterococcus faecalis* (*E. faecalis*), *Campylobacter jejuni* (*C. jejuni*), and more [168]. The antibacterial efficiency of PANI-coated fabrics reported in the literature is between 50 and 100% depending on the fabric

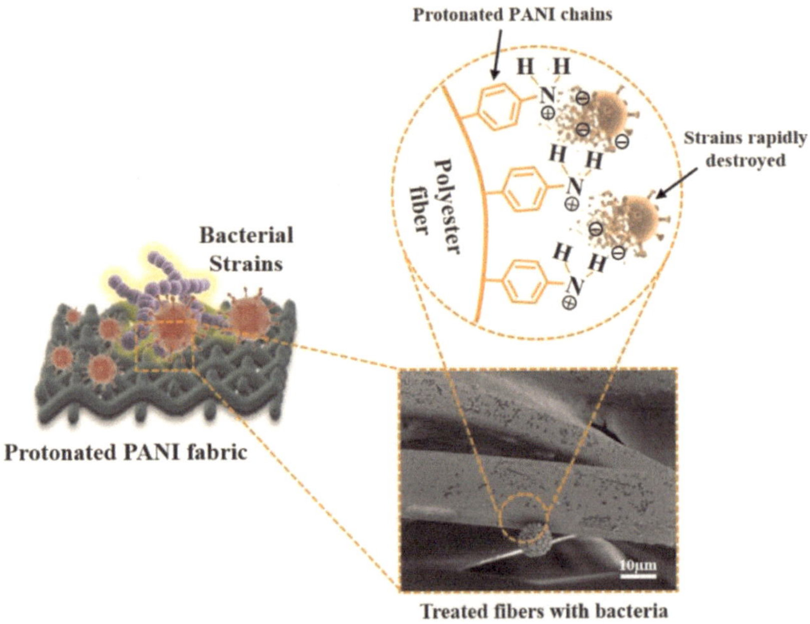

Figure 15.
Schematic representation of protonated polyaniline (PANI)-integrated fabric when exposed to the bacteria [52].

type, chemical composition of the coating, and the PANI fabrication method. **Table 6** summarizes some of the reported PANI-coated fabrics with antibacterial properties.

Amir's group applied PANI on non-woven fabrics, including viscose, PET, and PET:viscose (50:50), and counted the bacteria by a Colony count method that estimates the number of bacteria on the plate. The PET, viscose, and 50:50 PET:viscose PANI-coated fabrics exhibited a 100% antibacterial effect on the *S. aureus* population. In addition, PANI-coated viscose fabrics are very effective against *S. epidermidis*, showing a 100% reduction. Nevertheless, PANI-coated PET:viscose and PET fabrics showed a slightly more moderate effect against *S. epidermidis*, with a 99.65% and 99.9997% reduction in the bacterial population, respectively [42]. Mahat et al. fabricated a PANI/polyester fabric and analyzed the antibacterial properties using the Kirby-Bauer disc diffusion method. They measured the size of the inhibition zone (mm) around the tested fabric on Muellor Hinton (MH) agar, indicating the antibacterial activity level. They demonstrated that the Gram-positive bacteria group

Antibacterial fabric	Fabrication method	Test method	Bacteria type	Antibacterial efficiency	Ref.
PANI/viscose, PANI/PET, and PANI/PET:viscose (50:50)	In situ polymerization	Colony count	*S. aureus*	100%	[42]
			S. epidermidis	100%, 99.9997%, and 99.65%	
PANI/cotton	In situ polymerization	Colony count	*S. aureus*	95%	[170]
			E. coli	85%	
			Candida albicans	92%	

Antibacterial fabric	Fabrication method	Test method	Bacteria type	Antibacterial efficiency	Ref.
PANI/cotton and PANI/polyester	Dip-coating	Kirby-Bauer disc diffusion	*Klebsiella pneumoniae*	3.5 ± 0.170 mm and 10.3 ± 0.048 mm	[85]
			S. aureus	0 mm and 3.5 ± 0.170 mm	
			E. coli	0 mm and 5.0 ± 0.140 mm	
PANI/polyester	Dip-coating	Kirby-Bauer disc diffusion	*S. aureus*	22.30 ± 0.03 mm	[52]
			Staphylococcus aureus	21.50 ± 0.09 mm	
			S. epidermidis	19.11 ± 0.02 mm	
			P. aeruginosa	24.33 ± 0.02 mm	
			E. coli	14.12 ± 0.07 mm	
			S. typhi	21.35 ± 0.08 mm	
PANI/GO/cotton	Spray-coating	Kirby-Bauer disc diffusion	*E. coli*	10.2 mm	[82]
			S. aureus	11.0 mm	
PANI/carbon nitride (CN)/cotton	In situ polymerization	Colony count	*E. coli bacteria*	72.6%	[171]
PANI/Ag/cotton	In situ polymerization	Kirby-Bauer disc diffusion	*S. aureus*	3 mm	[106]
			E. coli	5 mm	
PANI/silk and PANI/copper sulfide nanoparticles (CuSNPs)/silk	In situ polymerization	Colony count	*E. coli*	45.3% and 99.64%	[60]
			S. aureus	~90% and 99.99%	
PANI/wool and PANI/chitosan/wool	Two-step polymerization	Colony count	*E. coli*	64.92% and 99.99%	[103]
			S. aureus	69.27% and 100%	

Table 6.
Antibacterial fabrics based on polyaniline (PANI).

exhibited a 19–22 mm diameter, while the Gram-negative bacterial groups exhibited a 14–24 mm diameter for the zone diameter breakpoint measurement. These results demonstrate a remarkable antibacterial performance of PANI-coated fabric against Gram-positive and Gram-negative bacteria [52]. Srinivasan et al. developed a carbon nitride (CN)/PANI-coated cotton fabric and tested its antibacterial effect on *E. coli*, indicating a 72.6% efficiency [171]. Stejskal et al. improved the antibacterial performance by adding Ag particles to the PANI-coated cotton fabric, displaying an antibacterial activity against *S. aureus* of 3 mm and *E. coli* of 5 mm [106]. In another research, Wei et al. tested the antibacterial properties of PANI/wool and PANI/chitosan (CTS)/wool fabrics against *E. coli*, showing 64.92% and 99.99% efficiency, respectively. Examination of the treated fabrics against *S. aureus* indicated a 69.27% and 100% efficiency for PANI/wool and PANI/CTS/wool, respectively. In the case of PANI/CTS/wool fabric, the antibacterial effect against *E. coli* and *S. aureus* was more than 99.99% even after 10 washing cycles [103]. **Table 6** summarizes the composition, fabrication methods, and antibacterial characteristics of the main reported antibacterial-based textiles.

4. Conclusions and future outlook

The increasing number of publications describing fabrication and properties of PANI-based fabrics reflects the tremendous potential of this exceptional material for a broad range of smart textile applications. Its processability from aqueous solutions is a significant feature behind exploring different coating and printing techniques toward the realization of large-scale production. In situ polymerization method is one of the most studied coating techniques that was extensively studied, providing several solutions for thickness and quality control of PANI coating on different types of fabrics. The results of the published studies indicate that the electrical conductivity of PANI-coated textiles can be modified within a wide range by controlling the polymerization conditions, performing the coating method, and using additional conductive fillers. This allows one to adjust the electrical and other properties of the coated fabrics based on the target application requirements. The unique combination of reversible stretchability of the textile yarns and PANI molecular chains is a key feature behind the pressure and strain-sensing ability of PANI-coated fabrics. Despite much of the progress that has been achieved in this field, future research should aim to increase the robustness of the coatings by establishing strong adhesion between the PANI coating and the fabric allowing multiple deformations of the fabric without damaging the PANI layer. The reported high values of EMI shielding effectiveness, combined with the intrinsic flexibility and drapability of the fabrics, indicate a strong potential of PANI-coated textile for protecting applications. The chemical structure of PANI is fundamental to the fabrication of smart textile sensors that respond to external stimuli by changing the electrical conductivity and optical properties. Furthermore, the antibacterial properties, gas- and pH-sensing abilities of PANI-coated fabrics can be reversible due to the rapid transformation between ES and EB forms. PANI establishes strong hydrogen bond interactions with cellulose-based fabrics containing hydroxyl groups on the surface, providing long-term coating stability. Nevertheless, the adhesion of PANI layer is weaker for low surface energy fabrics, such as polyester or nylon. Despite the several strategies introduced to improve coating adhesion by incorporating intermediate layers with polar chemical groups, further research is required to develop durable, functional coatings with high washing stability.

Overall, the research studies reviewed in this chapter provide a solid platform for the future development of PANI-coated fabrics with advanced and multifunctional properties and commercial realization.

Author details

Lihi Abilevitch, Limor Mizrahi, Gali Cohen, Shmuel Kenig and Elizabeth Amir*
Department of Polymer Materials Engineering, Shenkar College, Ramat-Gan, Israel

*Address all correspondence to: eamir@shenkar.ac.il

References

[1] Dan Y, Buzhor M, Raichman D, Menashe E, Rachmani O, Amir E. Covalent surface functionalization of nonwoven fabrics with controlled hydrophobicity, water absorption, and pH regulation properties. Journal of Applied Polymer Science. 2020;**138**(6): 1-11. DOI: 10.1002/app.49820

[2] Xu F et al. 3D arch-structured and machine-knitted triboelectric fabrics as self-powered strain sensors of smart textiles. Nano Energy. 2023;**109**:1-10. DOI: 10.1016/j.nanoen.2023.108312

[3] Lorusso E, Ali W, Hildebrandt M, Mayer-Gall T, Gutmann JS. Hydrogel functionalized polyester fabrics by UV-induced photopolymerization. Polymers (Basel). 2019;**11**(8):1-10. DOI: 10.3390/polym11081329

[4] Goswami S et al. Human-motion interactive energy harvester based on polyaniline functionalized textile fibers following metal/polymer mechano-responsive charge transfer mechanism. Nano Energy. 2019;**60**:794-801. DOI: 10.1016/j.nanoen.2019.04.012

[5] Ramlow H, Andrade KL, Immich APS. Smart textiles: An overview of recent progress on chromic textiles. Journal of the Textile Institute. 2021;**112**(1):152-171. DOI: 10.1080/00405000.2020.1785071

[6] Shah MA, Pirzada BM, Price G, Shibiru AL, Qurashi A. Applications of nanotechnology in smart textile industry: A critical review. Journal of Advanced Research. 2022;**38**:55-75. DOI: 10.1016/j.jare.2022.01.008

[7] Xue P, Tao X, Leung MY, Zhang H. Electromechanical properties of conductive fibres, yarns and fabrics. Wearable Electronics and Photonics. 2005;**2005**:81-104. DOI: 10.1533/9781845690441.81

[8] Júnior HLO, Neves RM, Monticeli FM, Dall Agnol L. Smart fabric textiles: Recent advances and challenges. Text. 2022;**2**(4):582-605. DOI: 10.3390/textiles2040034

[9] Dan Y, Popowski Y, Buzhor M, Menashe E, Rachmani O, Amir E. Covalent surface modification of cellulose-based textiles for oil-water separation applications. Industrial and Engineering Chemistry Research. 2020; **59**(13):5456-5465. DOI: 10.1021/acs.iecr.9b05785

[10] Hu J. Controlled release of hydrogel modified textile products. Journal of Controlled Release. 2011;**152**(Suppl):e31-e33. DOI: 10.1016/j.jconrel.2011.08.104

[11] Subbiah DK, Babu KJ, Das A, Rayappan JBB. NiOx nanoflower modified cotton fabric for UV filter and gas sensing applications. ACS Applied Materials & Interfaces. 2019;**11**(22): 20045-20055. DOI: 10.1021/acsami.9b04682

[12] Hartman C, Popowski Y, Raichman D, Amir E. Biodegradable polymer coating for controlled release of hydrophobic functional molecules from cotton fabrics. Journal of Coating Technology and Research. 2020;**17**(3): 669-679. DOI: 10.1007/s11998-019-00278-3

[13] Mocioiu AM, Mocioiu OC. Intelligent textile with doped polyaniline. Macromolecular Symposia. 2021;**396**(1):10-12. DOI: 10.1002/masy.202000331

[14] Hashemikia S, Hemmatinejad N, Ahmadi E, Montazer M. A novel cotton

fabric with anti-bacterial and drug delivery properties using SBA-15-NH2/ polysiloxane hybrid containing tetracycline. Materials Science and Engineering C. 2016;**59**:429-437. DOI: 10.1016/j.msec.2015.09.092

[15] Civan L, Kurama S. A review: Preparation of functionalised materials/ smart fabrics that exhibit thermochromic behaviour. Materials Science and Technology. 2021;**37**(18): 1405-1420. DOI: 10.1080/ 02670836.2021.2015844

[16] Wu Y et al. Long-term antibacterial protected cotton fabric coating by controlled release of chlorhexidine gluconate from halloysite nanotubes. RSC Advances. 2017;**7**(31):18917-18925. DOI: 10.1039/c7ra01464c

[17] Ballottin D et al. Antimicrobial textiles: Biogenic silver nanoparticles against Candida and Xanthomonas. Materials Science and Engineering C. 2017;**75**:582-589. DOI: 10.1016/j. msec.2017.02.110

[18] Ojstršek A et al. Metallisation of textiles and protection of conductive layers: An overview of application techniques. Sensors. 2021;**21**(10):3508. DOI: 10.3390/s21103508

[19] Kazani I et al. About the collinear four-point probe technique's inability to measure the resistivity of anisotropic electroconductive fabrics. Textile Research Journal. 2013;**83**(15): 1587-1593. DOI: 10.1177/ 0040517512452951

[20] Rehman A, Houshyar S, Reineck P, Padhye R, Wang X, Houshyar S. Multifunctional smart fabrics through nanodiamond-polyaniline nanocomposites. ACS Applied Polymer Materials. 2020;**2**(11):4848-4855. DOI: 10.1021/acsapm.0c00789

[21] Ojstršek A, Jug L, Plohl O. A review of electro conductive textiles utilizing the dip-coating technique: Their functionality, durability and sustainability. Polymers (Basel). 2022;**14** (21):4713. DOI: 10.3390/polym14214713

[22] Pattanarat K, Petchsang N, Osotchan T, Tang IM, Kim YH, Jaisutti R. High conductivity and durability textile gas sensor-based polyaniline-decorated-poly(3,4-ethylenedioxythiophene)/poly(4-styrenesulfonate) for ammonia detection. ACS Applied Polymer Materials. 2022;**4**(12):9006-9014. DOI: 10.1021/acsapm.2c01374

[23] Teli M, Dash S, Desai P. Polyaniline based conductive textiles. Journal of The Institution of Engineers (India): Series E. 2014;**95**(2):75-79. DOI: 10.1007/ s40034-014-0037-x

[24] Abu Sadat MS, Teay SH, Shahariar H, Fink PL, Albarbar A. Review on smart electro-clothing systems (SeCSs). Sensors. 2020;**20**(587): 1-23. DOI: 10.3390/s20030587

[25] Terse-Thakoor T et al. Thread-based multiplexed sensor patch for real-time sweat monitoring. npj Flexible Electronics. 2020;**4**(1):1-10. DOI: 10.1038/s41528-020-00081-w

[26] Krifa M. Electrically conductive textile materials—Application in flexible sensors and antennas. Text. 2021;**1**(2): 239-257. DOI: 10.3390/textiles1020012

[27] Moretti C, Tao X, Koncar V, Koehl L. A study on electrical performances and lifetime of a flexible electrochromic textile device. Autex Research Journal. 2014;**14**(2):76-81. DOI: 10.2478/aut-2014-0003

[28] Huang G, Liu L, Wang R, Zhang J, Sun X, Peng H. Smart color-changing

textile with high contrast based on a single-sided conductive fabric. Journal of Materials Chemistry C Materials. 2016; **4**(32):7589-7594. DOI: 10.1039/c6tc02051h

[29] Wei Y, Wang X, Torah R, Tudor J. Dispenser printing of electrochromic display on textiles for creative applications. Electronics Letters. 2017; **53**(12):779-781. DOI: 10.1049/el.2017.0119

[30] Shirakawa H, Louis EJ, MacDiarmid AG, Chiang CK, Heeger AJ. Synthesis of electrically conducting organic polymers: Halogen derivatives of polyacetylene, (CH)x. Journal of the Chemical Society, Chemical Communications. 1977;**16**:578-580. DOI: 10.1039/C39770000578

[31] Abu-Thabit NY. Chemical oxidative polymerization of polyaniline: A practical approach for preparation of smart conductive textiles. Journal of Chemical Education. 2016;**93**(9): 1606-1611. DOI: 10.1021/acs.jchemed.6b00060

[32] Nezakati T, Seifalian A, Tan A, Seifalian AM. Conductive polymers: Opportunities and challenges in biomedical applications. Chemical Reviews. 2018;**118**(14):6766-6843. DOI: 10.1021/acs.chemrev.6b00275

[33] Cohen David N et al. Electro-conductive fabrics based on dip coating of cotton in poly(3-hexylthiophene). Polymers for Advanced Technologies. 2017;**28**(5):583-589. DOI: 10.1002/pat.3857

[34] Islam GMN, Ali A, Collie S. Textile sensors for wearable applications: A comprehensive review. Cellulose. 2020; **27**(11):6103-6131. DOI: 10.1007/s10570-020-03215-5

[35] Stejskal J, Riede A, Hlavat D, Proke J, Helmstedt M, Holler P. The effect of polymerization temperature on molecular weight, crystallinity, and electrical conductivity of polyaniline. Synthetic Metal. 1998;**96**:55-61

[36] Yılmaz F, Küçükyavuz Z. The influence of polymerization temperature on structure and properties of polyaniline. E-Polymers. 2009;**5**:1-10

[37] Adams PN, Bowman D, Brown L, Yang D, Mattes BR. Molecular weight dependence of the physical properties of protonated polyaniline films and fibers. SPIE Smart Structures and Materials + Nondestructive Evaluation and Health Monitoring. 2001;**4329**:475-481. DOI: 10.1117/12.432679

[38] Macdiarmid AG, Epstein AJ. Polyaniline: Interrelationships between molecular weight, morphology, Donnan potential and conductivity. MRS Proceedings. 1992;**247**(1):565-576. DOI: 10.1557/PROC-247-565/METRICS

[39] Rai R, Roether JA, Boccaccini AR. Polyaniline based polymers in tissue engineering applications: A review. Progress in Biomedical Engineering. 2022;**4**:1-38. DOI: DOI 10.1088/2516-1091/ac93d3

[40] Beygisangchin M, Rashid SA, Shafie S, Sadrolhosseini AR. Preparations, Properties, and applications of polyaniline and polyaniline thin films—A review. Polymers (Basel). 2021;**13**:1-46

[41] Bhadra S, Khastgir D, Singha NK, Lee JH. Progress in preparation, processing and applications of polyaniline. Progress in Polymer Science (Oxford). 2009;**34**(8):783-810. DOI: 10.1016/j.progpolymsci.2009.04.003

[42] Jarach N et al. Hybrid antibacterial and electro-conductive coating for

textiles based on cationic conjugated polymer. Polymers (Basel). 2020;**12**(7): 1-15. DOI: 10.3390/polym12071517

[43] Shoaie N et al. Electrochemical sensors and biosensors based on the use of polyaniline and its nanocomposites: A review on recent advances. Microchimica Acta. 2019;**186**(7):465. DOI: 10.1007/s00604-019-3588-1

[44] Tissera ND, Wijesena RN, Rathnayake S, de Silva RM, de Silva KMN. Heterogeneous in situ polymerization of polyaniline (PANI) nanofibers on cotton textiles: Improved electrical conductivity, electrical switching, and tuning properties. Carbohydrates Polymers. 2018;**186**: 35-44. DOI: 10.1016/j.carbpol.2018.01.027

[45] Akbar SA, Satria E. Uv-vis study on polyaniline degradation at different phs and the potential application for acid-base indicator. Rasayan Journal of Chemistry. 2019;**12**(3):1212-1218. DOI: 10.31788/RJC.2019.1235370

[46] Gul S, Shah AUHA, Bilal S. Synthesis and characterization of processable polyaniline salts. Journal of Physics Conference Series. 2013;**439**(1):1-10. DOI: 10.1088/1742-6596/439/1/012002

[47] Roslan NC et al. Morphological and conductivity studies of polyaniline fabric doped phosphoric acid. Malaysian Journal of Analytical Sciences. 2020; **24**(5):698-706

[48] Paschoalin RT, Steffens C, Manzoli A, Paris EC, Herrmann PSP. PANI conductivity: A dependence of the chemical synthesis temperature. Macromolecular Symposia. 2012;**319**(1): 48-53. DOI: 10.1002/masy.201100242

[49] Stejskal J, Kratochvíl P, Radhakrishnan N. Polyaniline dispersions 2. UV-Vis absorption spectra. Synthetic Metals. 1993;**61**(3):225-231. DOI: 10.1016/0379-6779(93)91266-5

[50] Xing X et al. Twistable and tailorable V_2O_5/PANI/GO nanocomposites textile for wearable ammonia sensing. Sensors Actuators B Chemie. 2022;**351**:130944. DOI: 10.1016/j.snb.2021.130944

[51] Li DY et al. Functional polyaniline/MXene/cotton fabrics with acid/alkali-responsive and tunable electromagnetic interference shielding performances. ACS Applied Materials & Interfaces. 2022;**14**(10):12703-12712. DOI: 10.1021/acsami.2c00797

[52] Aizamddin MF et al. Antibacterial performance of protonated polyaniline-integrated polyester fabrics. Polymers (Basel). 2022;**14**(13):1-17. DOI: 10.3390/polym14132617

[53] Gicevicius M, Cechanaviciute IA, Ramanavicius A. Electrochromic textile composites based on polyaniline-coated metallized conductive fabrics. Journal of the Electrochemical Society. 2020; **167**(15):155515. DOI: 10.1149/1945-7111/abb0f3

[54] Parvez MS, Rahman MM, Samykano M, Yeakub Ali M. Electrochemical characterization and joule heating performance of polyaniline incorporated cotton fabric. Physics and Chemistry of the Earth;**129**:103323. DOI: 10.1016/j.pce.2022.103323

[55] Allison L, Hoxie S, Andrew TL. Towards seamlessly-integrated textile electronics: Methods to coat fabrics and fibers with conducting polymers for electronic applications. Chemical Communications. 2017;**53**(53): 7182-7193. DOI: 10.1039/c7cc02592k

[56] Barani H, Miri A, Sheibani H. Comparative study of electrically

conductive cotton fabric prepared through the in situ synthesis of different conductive materials. Cellulose. 2021; **28**(10):6629-6649. DOI: 10.1007/s10570-021-03928-1

[57] Mu S, Xie H, Wang W, Yu D. Electroless silver plating on PET fabric initiated by in situ reduction of polyaniline. Applied Surface Science. 2015;**353**:608-614. DOI: 10.1016/j.apsusc.2015.06.126

[58] Ding X, Wang W, Wang Y, Xu R, Yu D. High-performance flexible electromagnetic shielding polyimide fabric prepared by nickel-tungsten-phosphorus electroless plating. Journal of Alloys and Compounds. 2019;**777**: 1265-1273. DOI: 10.1016/j.jallcom.2018.11.120

[59] Wu G et al. Polyaniline/graphene-functionalized flexible waste mask sensors for ammonia and volatile sulfur compound monitoring. ACS Applied Materials & Interfaces. 2022;**14**: 56056-56064. DOI: 10.1021/acsami.2c15443

[60] Ren Y et al. Multifunctional textile constructed via polyaniline-mediated copper sulfide nanoparticle growth for rapid photothermal antibacterial and antioxidation applications. ACS Applied Nano Materials. 2022;**6**:1212-1223. DOI: 10.1021/acsanm.2c04797

[61] Ke G, Chowdhury MH, Jin X, Li W. Fabrication and properties of polyaniline/ramie composite fabric based on in situ polymerization. Polymers and Polymer Composites. 2021;**29**:S914-S925. DOI: 10.1177/09673911211028398

[62] Wang Y, Zhang X, Cao J, Huang X, Zhang X. Multifunctional E-Textiles Based on Biological Phytic Acid-Doped Polyaniline/Protein Fabric Nanocomposites. Advanced Materials Technology. 2021;**6**(6):1-8. DOI: 10.1002/admt.202100003

[63] Hirase R, Shikata T, Shirai M. Selective formation of polyaniline on wool by chemical polymerization, using potassium iodate. Synthetic Metals. 2004;**146**(1):73-77. DOI: 10.1016/j.synthmet.2004.06.009

[64] Zhao X, Sun J, Liu Y, Wang Y. The preparation and performance of polyaniline coated polyester fabrics. Journal of the Textile Institute. 2021; **112**(10):1700-1707. DOI: 10.1080/00405000.2020.1839217

[65] Maity D, Kumar RTR. Polyaniline anchored MWCNTs on fabric for high performance wearable ammonia sensor. ACS Sensors. 2018;**3**(9):1822-1830. DOI: 10.1021/acssensors.8b00589

[66] Qi J, Xu X, Liu X, Lau KT. Fabrication of textile based conductometric polyaniline gas sensor. Sens Actuators B Chem. 2014;**202**: 732-740. DOI: 10.1016/j.snb.2014.05.138

[67] Wu B et al. Electrical switchability and dry-wash durability of conductive textiles. Scientific Reports. 2015;**5**:1-9. DOI: 10.1038/srep11255

[68] Agarwala S. Enabling new possibilities in smart textiles through printed electronics. In: 2019 IEEE 9th International Nanoelectronics Conferences, INEC 2019. 2019. pp. 1-6. DOI: 10.1109/INEC.2019.8853853

[69] Ali AE, Jeoti V, Stojanović GM. Fabric based printed-distributed battery for wearable e-textiles: A review. Science and Technology of Advanced Materials. 2021;**22**(1):772-793. DOI: 10.1080/14686996.2021.1962203

[70] Stoppa M, Chiolerio A. Wearable electronics and smart textiles: A critical review. Sensors (Switzerland). 2014; **14**(7):11957-11992. DOI: 10.3390/s140711957

[71] Gong X, Huang K, Wu YH, Zhang XS. Recent progress on screen-printed flexible sensors for human health monitoring. Sens Actuators A Physics. 2022;**345**:1-19. DOI: 10.1016/j.sna.2022.113821

[72] Roshni SB, Jayakrishnan MP, Mohanan P, Surendran KP. Design and fabrication of an E-shaped wearable textile antenna on PVB-coated hydrophobic polyester fabric. Smart Materials and Structures. 2017;**26**(10): 1-8. DOI: 10.1088/1361-665X/aa7c40

[73] Zhu J, Song W, Peng J, Yin Y, Xu B, Wang C. Microwave thermally expanded graphene/polyaniline conductive paste for elaborate conductive pattern and conductive polyester fabric fabrication via screen printing. Journal of Coating Technology and Research. 2022;**19**(2): 477-485. DOI: 10.1007/s11998-021-00530-9

[74] Wang Z, Wang W, Jiang Z, Yu D. Low temperature sintering nano-silver conductive ink printed on cotton fabric as printed electronics. Progress in Organic Coating. 2016;**101**:604-611. DOI: 10.1016/j.porgcoat.2016.08.019

[75] Zhang X, Wang Y, Fu D, Wang G, Wei H, Ma N. Photo-thermal converting polyaniline/ionic liquid inks for screen printing highly-sensitive flexible uncontacted thermal sensors. European Polymer Journal. 2021;**147**:1-8. DOI: 10.1016/j.eurpolymj.2021.110305

[76] Laffitte Y, Gray BL. Real-Time potentiometric pH-sensor using a screen-printable polyaniline composite on textiles. In: IEEE International Conference on Flexible and Printable Sensors and Systems. 2021. pp. 19-22. DOI: 10.1109/FLEPS51544.2021.9469705

[77] Laffitte Y, Gray BL. Potentiometric pH sensor based on flexible screen-printable polyaniline composite for textile-based microfluidic applications. Micromachines (Basel). 2022;**13**(9):1-19. DOI: 10.3390/mi13091376

[78] Loffredo F, Burrasca G, Quercia L, Della Sala D. Gas sensor devices obtained by ink-jet printing of polyaniline suspensions. Macromolecular Symposia. 2007;**247**:357-363. DOI: 10.1002/masy.200750141

[79] Tseghai GB, Malengier B, Fante KA, Nigusse AB, Van Langenhove L. Integration of conductive materials with textile structures, an overview. Sensors (Switzerland). 2020;**20**(23):1-28. DOI: 10.3390/s20236910

[80] Stempien Z et al. Inkjet printing of polypyrrole electroconductive layers based on direct inks freezing and their use in textile solid-state supercapacitors. Materials. 2021;**14**(13):1-20. DOI: 10.3390/ma14133577

[81] Stempien Z, Rybicki T, Rybicki E, Kozanecki M, Szynkowska MI. In-situ deposition of polyaniline and polypyrrole electroconductive layers on textile surfaces by the reactive ink-jet printing technique. Synthetic Metals. 2015;**202**:49-62. DOI: 10.1016/j.synthmet.2015.01.027

[82] Zeng F, Qin Z, Chen Y, Shan X. Constructing polyaniline nanowire arrays as efficient traps on graphene sheets to promote compound synergetic effect in the assembled coating for multifunctional protective cotton fabrics. Chemical Engineering Journal. 2021;**426**(February):130819. DOI: 10.1016/j.cej.2021.130819

[83] Tang X, Yan X. Dip-coating for fibrous materials: Mechanism, methods and applications. Journal of Solgel Science and Technology. 2016;**81**(2): 378-404. DOI: 10.1007/s10971-016-4197-7

[84] Zeng W, Shu L, Li Q, Chen S, Wang F, Tao XM. Fiber-based wearable electronics: A review of materials, fabrication, devices, and applications. Advanced Materials. 2014;**26**: 5310-5336. DOI: 10.1002/adma.2014 00633

[85] Omar SNI et al. Electrically conductive fabric coated with polyaniline: Physicochemical characterisation and antibacterial assessment. Emergent Mater. 2020;**3**(4): 469-477. DOI: 10.1007/s42247-019-00062-4

[86] Amirabad R, Ramazani Saadatabadi A, Siadati MH. Preparation of polyaniline/graphene coated wearable thermoelectric fabric using ultrasonic-assisted dip-coating method. Materials Renewable Sustainable Energy. 2020; **9**(4):1-12. DOI: 10.1007/s40243-020-00181-7

[87] Alamer FA. Structural and electrical properties of conductive cotton fabrics coated with the composite polyaniline/ carbon black. Cellulose. 2018;**25**(3): 2075-2082. DOI: 10.1007/s10570-018-1667-9

[88] Elmogahzy YE. Performance characteristics of technical textiles: Part I: E-textiles. Engineering Text. 2020; **2020**:347-364. DOI: 10.1016/b978-0-08-102488-1.00014-9

[89] Pomfret SJ, Adams PN, Comfort NP, Monkman AP. Electrical and mechanical properties of polyaniline fibres produced by a one-step wet spinning process. Polymer (Guildf). 2000;**41**(6): 2265-2269. DOI: 10.1016/S0032-3861 (99)00365-1

[90] Perumalraj R. Electrical surface resistivity of polyaniline coated woven fabrics. Journal of Textiles Science and Engineering. 2015;**05**(03):1-5. DOI: 10.4172/2165-8064.1000196

[91] Neelakandan R, Madhusoothanan M. Electrical resistivity studies on polyaniline coated polyester fabrics. Journal of Engineering and Fiber Fabrication. 2010;**5**(3):25-29. DOI: 10.1177/155892501000500304

[92] Zhou X, Hu C, Lin X, Han X, Zhao X, Hong J. Polyaniline-coated cotton knitted fabric for body motion monitoring. Sensor Actuators A Physica. 2021;**321**:112591. DOI: 10.1016/j. sna.2021.112591

[93] Cao J, Wang C. Highly conductive and flexible silk fabric via electrostatic self assemble between reduced graphene oxide and polyaniline. Organic Electronics. 2018;**55**(January):26-34. DOI: 10.1016/j.orgel.2017.12.016

[94] Goswami S, Nandy S, Fortunato E, Martins R. Polyaniline and its composites engineering: A class of multifunctional smart energy materials. Journal of Solid State Chemical. 2023;**317**(July):123679. DOI: 10.1016/j.jssc.2022.123679

[95] Liu Y, Zhao X. The preparation and performance of a polyaniline/graphene composite coated fabric. Journal of the Textile Institute. 2021;**112**(8):1258-1265. DOI: 10.1080/00405000.2020.1809908

[96] Cheedarala RK, Parvez AN, Ahn KK. Electric impulse spring-assisted contact separation mode triboelectric nanogenerator fabricated from polyaniline emeraldine salt and woven carbon fibers. Nano Energy. 2018;**53**

(May):362-372. DOI: 10.1016/j.nanoen.2018.08.066

[97] Bajgar V et al. Cotton fabric coated with conducting polymers and its application in monitoring of carnivorous plant response. Sensors. 2016;**16**(4):498. DOI: 10.3390/s16040498

[98] Tang X et al. Functionalization of cotton fabric with graphene oxide nanosheet and polyaniline for conductive and UV blocking properties. Synthetic Metals. 2015;**202**: 82-88. DOI: 10.1016/j.synthmet.2015.01.017

[99] Attia RM, Yousif NM, Helal RH, Ali NM. Fabricating electronic textile using nano MnO2/polyaniline composites for capacitor device. Journal of Industrial Textiles. 2022;**51**(7): 1161-1180. DOI: 10.1177/1528083719896765

[100] Jin LN et al. High-performance textile supercapacitor electrode materials enhanced with three-dimensional carbon nanotubes/graphene conductive network and in situ polymerized polyaniline. Electrochimica Acta. 2017; **249**:387-394. DOI: 10.1016/j.electacta.2017.08.035

[101] Mocioiu AM, Tudor IA, Mocioiu OC. Application of polyaniline for flexible semiconductors. Coatings. 2021;**11**(1):1-10. DOI: 10.3390/coatings11010049

[102] He J, Li R, Gu F. Preparation of polyaniline/nylon conducting fabric by layer-by-layer assembly method. Journal of Applied Polymer Science. 2012; **128**(3):1673-1679. DOI: 10.1002/app.38331

[103] Yu J, Pang Z, Zhang J, Zhou H, Wei Q. Conductivity and antibacterial properties of wool fabrics finished by polyaniline/chitosan. Colloids and Surfaces A: Physicochemical and Engineering Aspects. 2018;**548**:117-124. DOI: 10.1016/j.colsurfa.2018.03.065

[104] Wu J, Zhou D, Looney MG, Waters PJ, Wallace GG, Too CO. A molecular template approach to integration of polyaniline into textiles. Synthetic Metals. 2009;**159**(12): 1135-1140. DOI: 10.1016/j.synthmet.2009.01.054

[105] Oh KW, Hong KH, Kim SH. Electrically conductive textiles by in situ polymerization of aniline. Journal of Applied Polymer Science. 1999;**74**(8): 2094-2101. DOI: 10.1002/(SICI)1097-4628(19991121)74:8<2094::AID-APP26>3.0.CO;2-9

[106] Maráková N et al. Antimicrobial activity and cytotoxicity of cotton fabric coated with conducting polymers, polyaniline or polypyrrole, and with deposited silver nanoparticles. Applied Surface Science. 2017;**396**:169-176. DOI: 10.1016/j.apsusc.2016.11.024

[107] Castano LM, Flatau AB. Smart fabric sensors and e-textile technologies: A review. Smart Materials and Structures. 2014;**23**(5):1-27. DOI: 10.1088/0964-1726/23/5/053001

[108] Guo L, Bashir T, Bresky E, Persson NK. Electroconductive textiles and textile-based electromechanical sensors-integration in as an approach for smart textiles. In: Vladan K, editor. Smart Textiles and Their Applications. UK: Woodhead Publishing, Elsevier Ltd; 2016. pp. 657-693. DOI: 10.1016/B978-0-08-100574-3.00028-X

[109] Bag A, Lee NE. Recent advancements in development of wearable gas sensors. Advanced Materials Technology. 2021;**6**(3):1-37. DOI: 10.1002/ADMT.202000883

[110] Fratoddi I, Venditti I, Cametti C, Russo MV. Chemiresistive polyaniline-based gas sensors: A mini review. Sensors Actuators B Chemical. 2015;**220**: 534-548. DOI: 10.1016/J. SNB.2015.05.107

[111] Alrammouz R, Podlecki J, Abboud P, Sorli B, Habchi R. A review on flexible gas sensors: From materials to devices. Sensors Actuators A Physics. 2018;**284**: 209-231. DOI: 10.1016/J. SNA.2018.10.036

[112] Kukla AL, Shirshov YM, Piletsky SA. Ammonia sensors based on sensitive polyaniline films. Sensors Actuators B Chemical. 1996;**37**(3):135-140. DOI: 10.1016/S0925-4005(97)80128-1

[113] Myat Swe M, Eamsa-Ard T, Kerdcharoen T. Fabrication of polyaniline coated conductive cotton for ammonia gas detection. MATEC Web of Conferences. 2018;**192**(02036):1-4. DOI: 10.1051/MATECCONF/201819202036

[114] Stempien Z et al. Ammonia gas sensors ink-jet printed on textile substrates. IEEE Sensors. 2017;**2017**:2-4. DOI: 10.1109/ICSENS.2016.7808457

[115] Collins GE, Buckley LJ. Conductive polymer-coated fabrics for chemical sensing. Synthetic Metals. 1996;**78**(2): 93-101. DOI: 10.1016/0379-6779(96) 80108-1

[116] Vivaldi F et al. Recent advances in optical, electrochemical and field effect pH sensors. Chem. 2021;**9**(2):1-17. DOI: 10.3390/chemosensors9020033

[117] Zhang W et al. Mechanical, electromagnetic shielding and gas sensing properties of flexible cotton fiber/polyaniline composites. Composites Science and Technology. 2020;**188**:1-7. DOI: 10.1016/j. compscitech.2019.107966

[118] de Almeida TM, da Silveira Maranhão F, de Carvalho FV, Middea A, de Araujo JR, de Souza Júnior FG. H_2S sensing material based on cotton fabrics modified with polyaniline. Macromolecular Symposia. 2018;**381**(1): 1-10. DOI: 10.1002/masy.201800111

[119] He M, Xie L, Luo G, Li Z, Wright J, Zhu Z. Flexible fabric gas sensors based on PANI/WO3 p−n heterojunction for high performance NH3 detection at room temperature. Science China Materials. 2020;**63**(10):2028-2039. DOI: 10.1007/s40843-020-1364-4

[120] Tang X et al. A low-cost polyaniline@textile-based multifunctional sensor for simultaneously detecting tactile and olfactory stimuli. Macromolecular Materials and Engineering. 2018;**303** (12):1-6. DOI: 10.1002/ mame.201800340

[121] P. Cavallo *et al.*, "Functionalized polyanilines made by nucleophilic addition reaction, applied in gas sensors field," Synthetic Metals, vol. 215, pp. 127–133, May 2016, doi: 10.1016/J. SYNTHMET.2016.02.013.

[122] Cavallo P, Acevedo DF, Fuertes MC, Soler-Illia GJAA, Barbero CA. Understanding the sensing mechanism of polyaniline resistive sensors. Effect of humidity on sensing of organic volatiles. Sensors Actuators B Chemical. 2015;**210**:574-580. DOI: 10.1016/j. snb.2015.01.029

[123] Bai H, Shi G. Gas sensors based on conducting polymers. Sensors. 2007;7: 267-307. DOI: 10.3390/s7030267

[124] Li ZF, Blum FD, Bertino MF, Kim CS. Understanding the response of nanostructured polyaniline gas sensors. Sensor Actuators B Chemical. 2013;**183**: 419-427. DOI: 10.1016/j.snb.2013.03.125

[125] Hong KH, Oh KW, Kang TJ. Polyaniline-Nylon 6 composite fabric for ammonia gas sensor. Journal of Applied Polymer Science. 2004;**92**(1):37-42. DOI: 10.1002/app.13633

[126] Adhikari B, Majumdar S. Polymers in sensor applications. Progress in Polymer Science (Oxford). 2004;**29**(7): 699-766. DOI: 10.1016/j. progpolymsci.2004.03.002

[127] Tang Y et al. Recent advances in wearable potentiometric pH sensors. Membranes (Basel). 2022;**12**(5):1-20. DOI: 10.3390/membranes12050504

[128] Hou X, Zhou Y, Liu Y, Wang L, Wang J. Coaxial electrospun flexible PANI//PU fibers as highly sensitive pH wearable sensor. Journal of Materials Science. 2020;**55**(33):16033-16047. DOI: 10.1007/s10853-020-05110-7

[129] Lindfors T, Ivaska A. pH sensitivity of polyaniline and its substituted derivatives. Journal of Electroanalytical Chemistry. 2002;**531**(1):43-52. DOI: 10.1016/S0022-0728(02)01005-7

[130] Choi MY et al. A fully textile-based skin pH sensor. Journal of Industrial Textiles. 2022;**51**:441S-457S. DOI: 10.1177/15280837211073361

[131] Guinovart T, Valdés-Ramírez G, Windmiller JR, Andrade FJ, Wang J. Bandage-based wearable potentiometric sensor for monitoring wound pH. Electroanalysis. 2014;**26**(6):1345-1353. DOI: 10.1002/elan.201300558

[132] Wu S. An overview of hierarchical design of textile-based sensor in wearable electronics. Crystals (Basel). 2022;**12**(555):1-32. DOI: 10.3390/ cryst12040555

[133] Dudem B, Mule AR, Patnam HR, Yu JS. Wearable and durable triboelectric nanogenerators via polyaniline coated cotton textiles as a movement sensor and self-powered system. Nano Energy. 2019;**55**:305-315. DOI: 10.1016/j. nanoen.2018.10.074

[134] Muthukumar N, Thilagavathi G, Kannaian T. Analysis of piezoresistive behavior of polyaniline-coated nylon Lycra fabrics for elbow angle measurement. Journal of the Textile Institute. 2017;**108**(2):233-238. DOI: 10.1080/00405000.2016.1161696

[135] Zhou Z et al. Textile-based mechanical sensors: A review. Materials. 2021;**14**(20):1-22. DOI: 10.3390/ ma14206073

[136] Callister W. Materials Science and Engineering: An Introduction. 7th ed. New York: John Wiley & Sons; 2007: 131-166

[137] Beer FP, Russell Johnston JE, Dewolf JT, Mazurek DF. Mechanics of Materials. 6th ed. New York, NY: McGraw-Hill. 2006

[138] Soloman S. Sensors Handbook. 1999

[139] Gonçalves C, da Silva AF, Gomes J, Simoes R. Wearable e-textile technologies: A review on sensors, actuators and control elements. Inventions. 2018;**3**(14):1-13. DOI: 10.3390/inventions3010014

[140] Qin QH. Advanced Mechanics of Piezoelectricity 2013. Berlin, Heidelberg: Springer Jointly published with Higher Education Press; DOI: 10.1007/978-3-642-29767-0

[141] Manbachi A, Cobbold RSC. Development and application of piezoelectric materials for ultrasound generation and detection. Ultrasound. 2011;**19**(4):187-196. DOI: 10.1258/ ult.2011.011027

[142] Chen X, Li B, Qiao Y, Lu Z. Preparing polypyrrole-coated stretchable textile via low-temperature interfacial polymerization for highly sensitive strain sensor. Micromachines (Basel). 2019;**10**(11):788. DOI: 10.3390/mi10110788

[143] Wu J, Zhou D, Too CO, Wallace GG. Conducting polymer coated lycra. Synthetic Metals. 2005;**155**:698-701. DOI: 10.1016/j.synthmet.2005.08.032

[144] Fan Q, Zhang X, Qin Z. Preparation of polyaniline/polyurethane fibers and their piezoresistive property. Journal of Macromolecular Science, Part B: Physics. 2012;**51**(4):736-746. DOI: 10.1080/00222348.2011.609795

[145] Muthukumar N, Thilagavathi G, Kannaian T. Design and development of polyaniline-coated fabric strain sensor for goniometry applications. International Journal of Science and Engineering Applications. 2013;**7560**:38-42. DOI: 10.7753/ijseancrtam.1010

[146] Rashid IA et al. Highly resilient and responsive fabric strain sensors: Their effective integration into textiles and wireless communication for wearable applications. Sensor Actuators A Physics. 2022;**346**:1-11. DOI: 10.1016/j.sna.2022.113836

[147] Huang Y, Gao L, Zhao Y, Guo X, Liu C, Liu P. Highly flexible fabric strain sensor based on graphene nanoplatelet–polyaniline nanocomposites for human gesture recognition. Journal of Applied Polymer Science. 2017;**134**(39):1-8. DOI: 10.1002/app.45340

[148] Tang X et al. Water-repellent flexible fabric strain sensor based on polyaniline/titanium dioxide-coated knit polyester fabric. Iranian Polymer Journal (English Edition). 2015;**24**(8):697-704. DOI: 10.1007/s13726-015-0361-0

[149] Liu K et al. Polyaniline Nanofiber Wrapped Fabric for High Performance Flexible Pressure Sensors. Polymers (Basel). 2019;**11**(1120):1-11. DOI: 10.3390/polym11071120

[150] Xu R et al. A flexible, conductive and simple pressure sensor prepared by electroless silver plated polyester fabric. Colloids and Surfaces A: Physicochemical and Engineering Aspects. 2019;**578**(February):123554. DOI: 10.1016/j.colsurfa.2019.04.096

[151] Ma Z, Wang W, Yu D. Assembled wearable mechanical sensor prepared based on cotton fabric. Journal of Materials Science. 2020;**55**(2):796-805. DOI: 10.1007/s10853-019-04035-0

[152] Celozzi S, Araneo R, Lovat G. Electromagnetic Shielding. Hoboken, New Jersey: John Wiley & Sons, Inc.; 2008. DOI: 10.1002/9780470268483

[153] Zubair K et al. Study of mechanical, electrical and EMI shielding properties of polymer-based nanocomposites incorporating polyaniline coated graphene nanoparticles. Nano Express. 2021;**2**(1):1-14. DOI: 10.1088/2632-959X/abe843

[154] Kim SH, Seong JH, Oh KW. Effect of dopant mixture on the conductivity and thermal stability of polyaniline/nomex conductive fabric. Journal of Applied Polymer Science. 2002;**83**(10):2245-2254. DOI: 10.1002/app.10211

[155] Onar N, Akşit AC, Ebeoglugil MF, Birlik I, Celik E, Ozdemir I. Structural, electrical, and electromagnetic properties of cotton fabrics coated with polyaniline and polypyrrole. Journal of Applied Polymer Science. 2009;**114**(4):2003-2010

[156] Zhou Z, Sun J. Fabrication and characterization of polyaniline on cotton

fabric via two-step in-situ polymerization. Journal of Physics Conference Series. 2021;**1790**(1):012070. DOI: 10.1088/1742-6596/1790/1/012070

[157] Muthukumar N, Thilagavathi G. Development and characterization of electrically conductive polyaniline coated fabrics. Indian Journal of Chemical Technology. 2012;**19**(6):434-441

[158] Muthukumar N, Thilagavathi G, Kannaian T. Polyaniline-coated nylon lycra fabrics for strain sensor and electromagnetic interference shielding applications. High Performance Polymers. 2015;**27**(1):105-111. DOI: 10.1177/0954008314540313

[159] Dhawan SK, Singh N, Venkatachalam S. Shielding effectiveness of conducting polyaniline coated fabrics at 101 GHz. Synthetic Metals. 2001;**125**(3):389-393. DOI: 10.1016/S0379-6779(01)00478-7

[160] Joseph N, Varghese J, Sebastian MT. In situ polymerized polyaniline nanofiber-based functional cotton and nylon fabrics as millimeter-wave absorbers. Polymer Journal. 2017;**49**(4):391-399. DOI: 10.1038/pj.2016.121

[161] Pan T, Zhang Y, Wang C, Gao H, Wen B, Yao B. Mulberry-like polyaniline-based flexible composite fabrics with effective electromagnetic shielding capability. Composite Science Technology. 2020;**188**(December 2019):107991. DOI: 10.1016/j.compscitech.2020.107991

[162] Gulati R, Sharma S, Sharma RK. Antimicrobial textile: Recent developments and functional perspective. Polymer Bulletin. 2022;**79**(8):5747-5771. DOI: 10.1007/s00289-021-03826-3

[163] Maruthapandi M, Saravanan A, Luong JHT, Gedanken A. Antimicrobial properties of the polyaniline composites against pseudomonas aeruginosa and klebsiella pneumoniae. Journal of Functional Biomaterials. 2020;**11**(3):59. DOI: 10.3390/jfb11030059

[164] Bhandari V, Jose S, Badanayak P, Sankaran A, Anandan V. Antimicrobial finishing of metals, metal oxides, and metal composites on textiles: A systematic review. Industrial and Engineering Chemistry Research. 2022; **61**:86-101. DOI: 10.1021/acs.iecr.1c04203

[165] Unango FJ, Ramasamy KM. A review on the investigation of biologically active natural compounds on cotton fabrics as an antibacterial textile finishing. International Research Journal of Science and Technology. 2019;**1**(1):49-55

[166] Tan L et al. A review of antimicrobial fabric containing nanostructures metal-based compound. Journal of Vinyl and Additive Technology. 2019;**25**:E3-E27. DOI: 10.1002/vnl.21606

[167] Qian J et al. Highly stable, antiviral, antibacterial cotton textiles via molecular engineering. Nature Nanotechnology. 2023;**18**(2):168-176. DOI: 10.1038/s41565-022-01278-y

[168] Robertson J, Gizdavic-Nikolaidis M, Nieuwoudt MK, Swift S. The antimicrobial action of polyaniline involves production of oxidative stress while functionalisation of polyaniline introduces additional mechanisms. PeerJ. 2018;**6**:1-36. DOI: 10.7717/peerj.5135

[169] Kucekova Z, Kasparkova V, Humpolicek P, Sevcikova P, Stejskal J.

Antibacterial properties of polyaniline-silver films. Chemical Papers. 2013;**67**(8):1103-1108. DOI: 10.2478/s11696-013-0385-x

[170] Seshadri DT, Bhat N. Use of polyaniline as an antimicrobial agent in textiles. Indian Journal of Fibre & Textile Research. 2005;**30**(2):204-206

[171] Meganathan P, Subbaiah S, Selvaraj LM, Subramanian V, Pitchaimuthu S, Srinivasan N. Photocatalytic self-cleaning and antibacterial activity of cotton fabric coated with polyaniline/carbon nitride composite for smart textile application. Phosphorus, Sulfur and Silicon and the Related Elements. 2022;**197**(3):244-253. DOI: 10.1080/10426507.2021.2012779

Chapter 5

PANI-Based Sensors: Synthesis and Application

Anita Grozdanov, Perica Paunović, Iva Dimitrievska and Aleksandar Petrovski

Abstract

In this chapter, we will present different methods of synthesis of PANI-based nanocomposites and their applications as bionanosensors, pH, and gas nanosensors. In this chapter, a comparison of various methods of synthesis of PANI-based nanocomposites with carbon nanotubes and graphene, as well as the production of nanosensors based on Screen Printed Electrodes will be given. Parallel, complete electrochemical and physical characterization of SPE-based nanosensor electrodes will be presented. For biosensing applications, various pharmaceutical active components will be reported. For pH testing, results of seawater testing in various parts of Europe (Sardinia, Barcelona, Napoli) will be reported. Gas-sensing analysis was done for SO_4, CO_2, and NH_3 gases.

Keywords: PANI-based sensors, electrochemical synthesis, pH sensing, gas sensing, PANI-based nanocomposites

1. Introduction

Electroconductive polymers are a class of modern materials particularly interesting to researchers because of their proven sensing ability [1]. These low-cost materials have demonstrated desired properties for developing room temperature operable gas sensors. Conductive polymers receive attention for their vast use in electrochemical applications, mainly for the improvement of electronics such as sensors, optoelectronic, and photonic devices [2].

Last decade, particular interest has been directed toward the design and development of chemical sensors based on electro-conductive polymers such as polyaniline (PANI). Especially, great interest was given to PANI-based composites due to the fact that composite materials made of conductive polymer or semi-conductors have been extensively used to improve the properties and sensing performance of gas sensing at room temperature [3]. In addition to electronic conductive polymers, nanomaterials are a crucial sensing component for the further development of gas sensors. Their ability to interact with the surroundings at a nanoscale level, large specific surface area, and high reactivity show the incredible performance of nanomaterials employed as excellent sensors with superior properties. Polymer-based electrodes can be implemented as an exceptional system for the detection of gaseous chemicals [4].

Polymer-based chemical sensors with specific functionalities can successfully recognize a particular compound within a mixture, providing rapid identification and qualitative analysis. Introducing a second component such as nanomaterials into PANI enhances its performance because of the synergetic effect between the two structures and thereby expands its application scope in the field of electronic devices.

Due to its p-type semiconductivity, polyaniline is characterized as a versatile electro polymer because of its excellent optical and electrical properties, conductivity, and ability to work at room temperature [3, 5]. Owing to the unique conduction mechanism, ease of synthesis, low cost, and high environmental stability, PANI is serving as a potential candidate for the fabrication of sensitive layers of novel gas sensors [6]. PANI's outstanding conductivity arises from its simple and reversible acid/base doping-dedoping chemistry. The doping process includes removal of electrons from the polymer's backbone, causing the cation radical to act as a charge carrier [7].

Not only PANI but also nano-sized PANI with high surface-to-volume ratio and unique electrical properties that can enhance the gas adsorption/desorption and thus promote the response and recovery processes is one of the ideal candidates for the development of PANI composites-based gas. The PANI/PVA composites have been reported for carbon dioxide sensing by Doan et al. [8]. PANI/TiO2 nanocomposites and PANI/sodium superoxide composite exhibited high sensitivity toward the CO2 sensor, and the fabricated sensor shows good response and recovery time [9]. The sensor performance of pure and PANI/Cu–ZnS composite were studied by Parangusan et al. [10]. Their results indicated that the PANI/Cu–ZnS composite exhibits a higher sensor response upon exposure to CO2 gas at room temperature. They discussed the enhanced sensing related to the p/n hetero-junction and porous microstructure of the core-shell structure. He et al. successfully produced a PANI/WO3@cotton thread-based flexible sensor that is capable of detecting NH3 at room temperature [11]. The WO3 nano-blocks were synthesized using a hydrothermal method, while the PANI/WO3@cotton thread was prepared by in-situ polymerization. The molar ratio of WO3 was varied and for 10% of WO3, the obtained sensor demonstrated the highest gas response upon exposure to 100 ppm NH3.

Shen et al. designed a wireless passive gas sensor based on alumina ceramic in which the sensor components were based on acidified CNT and PANI composites [12]. The gas sensing layer worked by physical adsorption of NH3 molecules. The wireless passive measurement was realized through the changes in resonant frequency. The relevant tests have shown that the sensor was able to detect a wide range of concentrations with a sensitivity of 0.04 MHz/ppm in the concentration of 300 ppm NH3. Parmar and his team worked on PANI and PANI/Graphene film-based sensors for the detection of the toluene [13]. The graphene–PANI ratio in the nanocomposite polymer film was optimized at 1:2. For the film's preparation they used N-methyl-2-pyrrolidone (NMP) solvent. The sensing behaviors of the films were analyzed at different temperatures (30, 50, and 100°C) for 100 ppm toluene in air. They found out that the nanocomposite Graphene-PANI films have exhibited better overall toluene sensing behavior in terms of sensor response and recovery time as well as repeatability [13].

Gaikwad et al. [14] reported the synthesis of PANI nanofibers and two nanocomposite systems: Polyaniline/Graphene Oxide (PANI/GO) and Polyaniline/Graphene Oxide/Zinc Oxide (PANI/GO/ZnO), used for sensing of NH3, LPG, CO_2 and H2S gases at room temperature. The authors observed better selectivity and sensitivity from all systems toward NH_3 at room temperature. PANI/GO/ZnO nanocomposite showed the best performance with a response of 5.706 for 1000 ppm NH_3 at around 80°C and a rapid recovery time of only 90 seconds. Polyaniline/zinc oxide (PANI/ZnO) hybrid film-based sensors have been developed for ammonia detection at room temperature by Zhu et al. [15]. The obtained sensor exhibited a p-type semiconductor behavior and better response than pristine PANI. ZnO nanorod arrays incorporated into the PANI create a nanoscale gap for gas diffusion and provide abundant adsorption sites, thus enhancing the sensors' response. The length of the nanorods is proportional to the sensitivity i.e., longer nanorods provide an efficient gap for gas diffusion, leading to better sensitivity.

Liu et al. [16] proposed an NH3 gas detector based on pristine and nanostructuralized PANI thin film, prepared by chemical oxidation polymerization and spin coating approach with further etching via reactive ion etching (RIE). The nanostructuralized PANI thin film sensor showed an increased response from 1.16 to 3.19 while increasing the NH3 concentration from 3 ppm to 990 ppm. The response, reproducibility, and selectivity toward NH3 showed superior properties of the nanostructuralized PANI film compared to the pristine PANI film sensor.

Macangnano et al. [17] investigated systems of PANI nanofibrous layers with different electrospinnable hosting polymers (polystyrene (PS), polyvinylpyrrolidone (PVP), and polyethylene oxide (PEO)), tested against traces of nitrogen oxide and ammonia. The authors reported good selectivity and rapid sensor response due to the components' both high porosity and high interaction surface. The hosting polymers are reported as modulators of the sensors' properties such as: PANI-PEO, a sensor with good electrical performance and high sensitivity to NH_3, but unstable while subjected to environmental stress; PANI-PVP, with good selectivity toward NO_2 but unstable while exposed to humidity; PANI-PS, stable and promising sensor, with good performance even when exposed to high humidity (up to 50% RH) and high temperature. Electrospun nanofibers have been confirmed as promising candidates for the development of gas sensors with high sensitivity, offering an improvement of the surface area-to-volume ratio. Korent et al. [18] described a novel and controllable method for the synthesis of PANI sensors using electrochemical deposition via cyclic voltammetry (CV) on golden screen-printed electrodes (SPEs), used for the detection of NH3. The proposed sensor showed good performances in terms of repeatability, reproducibility, and sensitivity in the range of 32–1100 ppb of NH3. Zhu et al. [19] developed a high-performance ammonia gas sensor based on self-assembly polyaniline films, prepared with the assistance of sodium dodecyl benzene sulfonate (SDBS). The SDS-functionalized PANI film-based sensor showed a good response toward NH_3 in the concentration range of 5.4–40 ppm NH_3, a low detection limit of 0.1 ppm, and a recovery time of 12 s. The surfactant functionalized PANI film enhances the gas sensing performance as a result of its structure ordered to accelerate the electron transport rate and the protonation/deprotonation properties.

Moreover, PANI is investigated in depth as an active component for other electrical devices such as pH sensors. Since solution pH has a crucial role in chemical reactions, exact determination and monitoring of pH are important in various fields [20].

In this chapter, we will discuss the synthesis and application of PANI, as one of the most promising conductive polymer substrates for the development and improvement of electrical devices such as sensors.

2. Synthesis and application of PANI-based nanocomposites

Polyaniline can be produced by chemical or electrochemical polymerization. Both include redox processes. Chemical polymerization is carried out in strongly acidic aqueous solutions in the presence of various oxidizing agents such as $K_2Cr_2O_7$,

$(NH)_4S_2O_8$, or KIO3 [21–23]. Its advantage is obtaining PANI to a large volume scale [23] compared to electropolymerization which is limited to dimensions of the electrode. Electrochemical polymerization is carried out in an aqueous solution of aniline in strong oxidizing protonic acid (mostly H_2SO_4) [24], using different electrochemical techniques such as: potentiostatic, galvanostatic, and cyclic potentiodynamic methods [25]. The product of electrolysis—polyaniline, is obtained on the anode (oxidative process). Advantages of the electrochemical route of polyaniline synthesis, especially for PANI-based composites and sensors are [26]: (1) good adhesion on the transducer surface (ex. screen printed electrodes), (2) control of layer thickness (controlled electron transfer with the electrode), (3) control of the morphology (using an appropriate electrolyte and its hydrodynamic regime), (4) no need for polymerization initiator or heating and (5) higher conductivity of the produced polymer/composite [27]. Depending on the applied potential, the mechanism of the PANI electropolymerization involves the exchange of electrons of the aniline with electrode forming different electronic states of the polyaniline such as leucoemeraldine (fully reduced), emeraldine (half-oxidized), and pernigraniline (fully oxidized) [28–30], where emeraldine is the highly conductive one. Electropolymerization of PANI was proved as an auto-catalyzed process [31].

The following text will explained our own procedure for the electrochemical synthesis of polyaniline and nanocomposite based on PANI, reinforced with carbon nanostructures (CNSs)—graphene(G) and multi-walled carbon nanotubes (MWCNTs) [32, 33]. The used graphene was obtained by molten salt electrolysis in the lab of the Faculty of Technology and Metallurgy in Skopje. Before the usage, the graphene was treated in a 10 wt% solution of H_2O_2 for 2 h and later, in a 40 wt% solution of HF for 1 h. MWCNTs were received from JRC (No.231, ISPRA, d = 10÷40 nm, purity ~94%) and used without additional treatment.

The first step in the determination of the electropolymerization conditions was cyclic voltammetry scanning of both systems—pure PANI and PANI + CNSs. The measurements were performed in three-electrode cells, where the working and counter electrodes were platinum tiles with a working surface of 10 cm^2, while as a reference one, a saturated calomel electrode (SCE) was used. Two types of electrolytes were prepared: 0.1 M aniline +0.5 M H2SO4 and 0.1 M aniline +0.5 M H_2SO_4 + CNSs. Before being dispersed into the electrolyte, CNSs were sonicated in an ultrasonic bath for 30 minutes. During the measurements, the electrolyte was stirred by a magnetic stirrer (200 rpm) at ambient temperature. The electrochemical set-up was consisted of potentiostat/galvanostat METROHM Autolab PGSTAT 128 N, three-electrode cells, and corresponding software for data acquisition (**Figure 1**). The potential range of scanning was from −0.2 to 1 V.

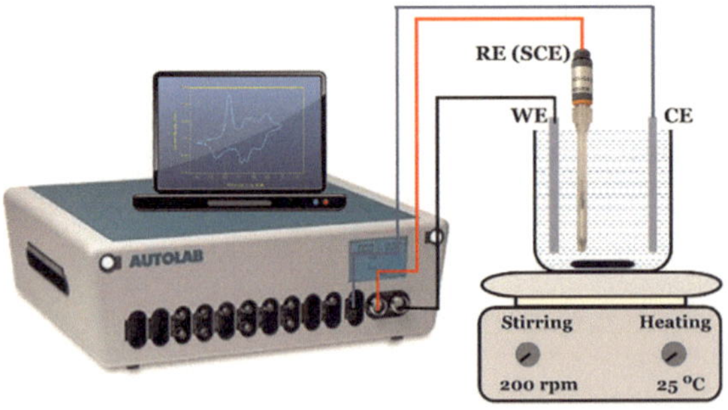

Figure 1.
Electrochemical set-up for cyclic voltammetry measurements.

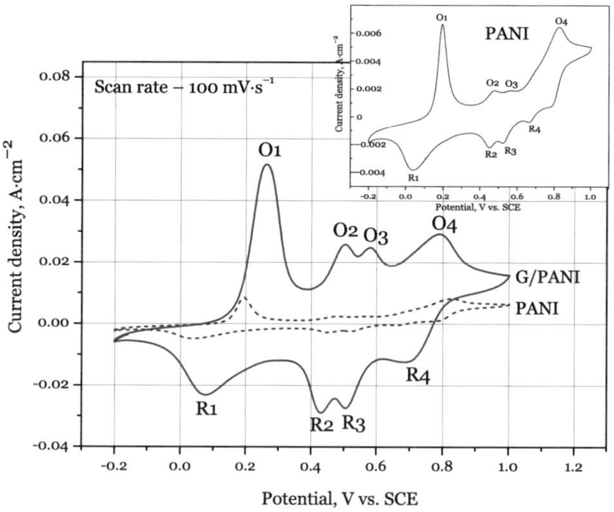

Figure 2.
Cyclic electrochemical spectra of electropolymerization of pure PANI (0.1 M aniline +0.5 M H2SO4) and G/PANI nanocomposite (0.1 M aniline +0.5 M H2SO4 + 3%wt. graphene related to aniline weight). In the inset, only the spectrum of electropolymerization of pure PANI is shown.

The electrochemical spectra of both studied systems are shown in **Figure 2**. Both spectra show similar redox peaks, but the spectrum of the G/PANI shows a much higher current response as a result of the presence of graphene with high electrical conductivity. So, there is more electron exchange and electrochemical processes occur faster in this system. Because redox peaks of the pure PANI system are not so pronounced, this spectrum is shown separately in the inset of **Figure 2**. Corresponding potential values of peaks appearance are summarized in **Table 1**.

The peak O1 in the anodic potential region corresponds to oxidation of the reduced form of PANI—leucoemeraldine to half-oxidized form—emeraldine [34–36]. The opposite peak in cathodic region R1, corresponds to the reverse reaction. According to the literature data [34, 37], this process occurs with the removal of electrons from the nitrogen atoms of the amine between the benzene rings, where the nitrogen atom acquires a cationic character. The middle redox pairs O2/R2 and O3/R3 can be ascribed to the oxidation/reduction of intermediate products [37, 38]. The oxidation peak O2 corresponds to the formation of benzoquinone, while the opposite cathodic peak R2 to its reduction to hydroquinone [35]. These redox reactions are the result of over-oxidation or degradation of PANI film [39]. O3/R3 is related to the formation of p-aminophenol/benzoquinoneimine [39, 40]. The last peak O4 denotes the oxidation of half-oxidized emeraldine to the fully oxidized pernigraniline form of PANI [34–36]. The opposite reaction is denoted with peak R4 in the cathodic region. Form the cyclic voltammogram, the potential region of formation of the partially oxidized emeraldine (the electroconductive form of PANI)

System	O1	O2	O3	O4	R1	R2	R3	R4
PANI	0.19	0.48	0.56	0.82	0.037	0.45	0.52	0.79
G/PANI	0.26	0.50	0.58	0.79	0.076	0.43	0.50	0.70

Table 1.
Potential position of the characteristic redox peaks for the studied system, from cyclic voltammograms in ***Figure 2****.*

can be estimated. According to the voltammogram it takes place in the potential region of 0.64–0.8 V. But, in this potential region is possible to obtain pernigraniline —a non-conductive form of PANI. In order to avoid this and to determine precisely the working potential of PANI electropolymerization, the steady-state polarization measurement was performed.

The steady-state change of the current in the potential region from 0.6 to 1.1 V is shown in **Figure 3**. As can be seen, the oxidation of emeraldine begins at 0.7 V vs. SCE and completed to pernigraniline at 0.9 V. It was expected that the formation of electroconductive PANI should occur in the middle of the oxidation region of emeraldine, i.e., at 0.8 V. But, the electropolymerization at this potential lead to the formation of a dark blue film of non-conductive pernigraniline. Therefore, the electropolymerization should be performed at a lower potential. The next attempt of electropolymerization at 0.75 V, lead to the formation of desirable conductive, green-colored emeraldine.

After the determination of the optimal electropolymerization potential, PANI and CNSs/PANI films were deposited on screen-printed electrodes (SPE) aimed for different sensing applications. Electropolymerization was performed at potentiostatic conditions, at 0.75 V, using an electrochemical set-up (potentiostat/galvanostat WENKING HC 500) as is shown in **Figure 4**.

During the potentiostatic electropolymerization (E = const.), it was observed that current density continuously increased, highlighting the autocatalytic character of the electropolymerization process. This is illustrated in **Figure 5**, where the change of current density during the time for electropolymerization of pure PANI and nanocomposites G/PANI and MWCNTs/PANI are shown. Increasing of the current density can be explained by the initial formation of the polymer/composite film, which possesses increased surface roughness related to the initial pure electrode surface [35, 41]. As the thickness of the polymer/composite film increases, surface roughness increases and consequently, current density increases. Initially, there is not so pronounced current increase. This can be ascribed to the induction period of the

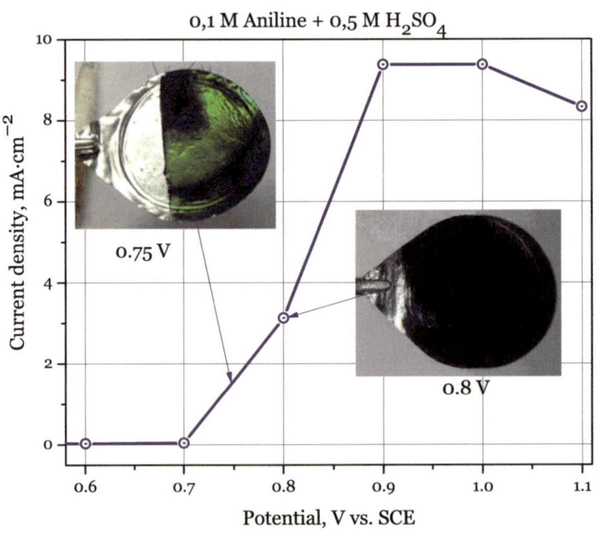

Figure 3.
Steady-state polarization curve in the system 0.1 M aniline +0.5 M H2SO4.

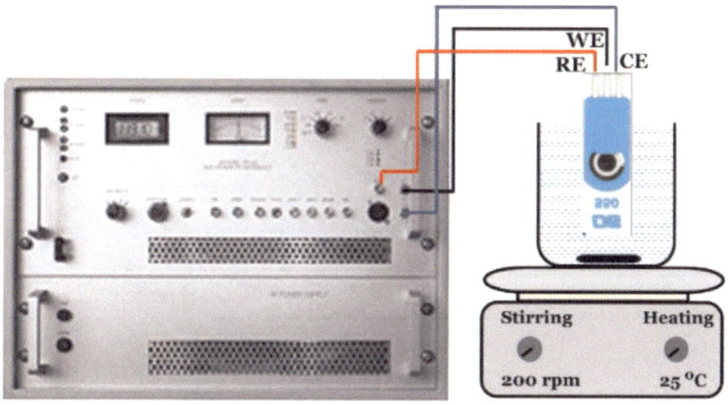

Figure 4.
Electrochemical set-up for electropolymerization of PANI on screen printed electrodes (SPE).

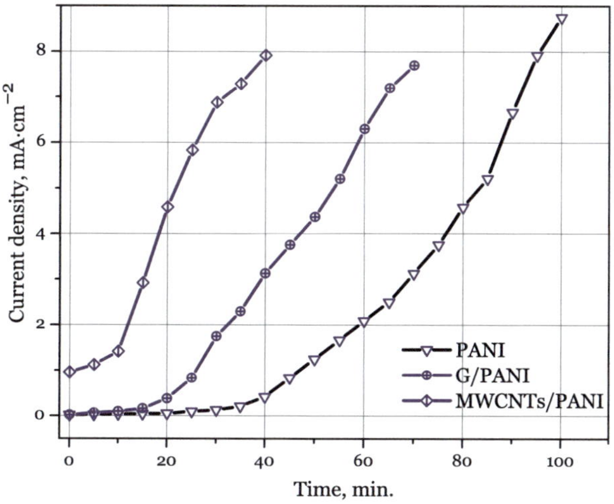

Figure 5.
Change of the current density with time during the potentiostatic electropolymerization of PANI, G/PANI, and MWCNTs/PANI, at 0.75 V vs. SCE.

polymer/composite film formation. The induction period consists of a few consequent processes: (1) oxidation of aniline to radical cations, (2) their polymerization to an oligomer of aniline, and (3) formation of PANI by autocatalytic reaction [42–44]. Another observation that we can highlight from **Figure 5** is that the electropolymerization of composites is considerably more intensive than that of pure PANI. This is the result of the much higher electrical conductivity of CNSs included in the polymer film, enabling faster electrons exchange of electrons and consequently, higher current density. Nanocomposite reinforced with MWCNTs has shown the highest current density of the electropolymerization process.

Cyclic voltammetry was also used for the determination of the double layer capacity of the obtained PANI and CNSs/PANI nanocomposite, using an electrochemical set-up as in **Figure 1**. The voltammograms were scanned at different scan rates: 10, 20, 50, and 100 mV·s^{-1} and shown in **Figure 6** for pure PANI and in **Figure 7** for G/PANI nanocomposite. Double layer capacity C_{dl} can be calculated by the following equation [45]:

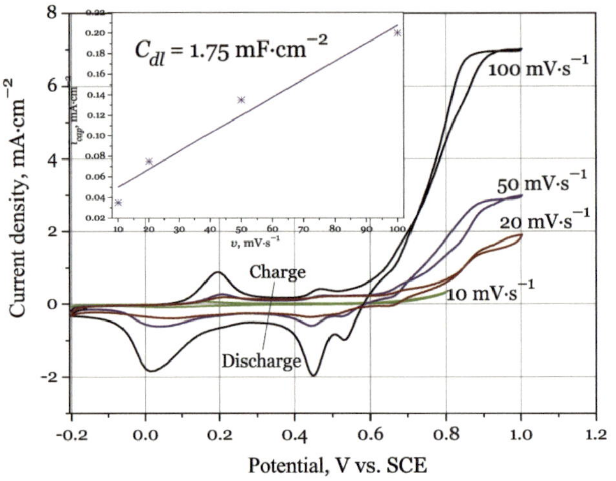

Figure 6.
Cyclic electrochemical spectra of pure PANI (0.1 M aniline +0.5 M H2SO4) at different scan rates: 10, 20, 50, and 100 mV·s⁻¹. In inset is shown the change of capacitance current density by the change of scan rate.

$$C_{dl} = \frac{di_{cap.}}{dv_i}, \tag{1}$$

where v_i is scan rate and $i_{cap.}$ is the capacitance current density at intersected potential in the region of double-layer charging and discharging as shown in **Figures 6** and 7. Capacitance current density is the arithmetic mean of the absolute values of anodic and cathodic current density:

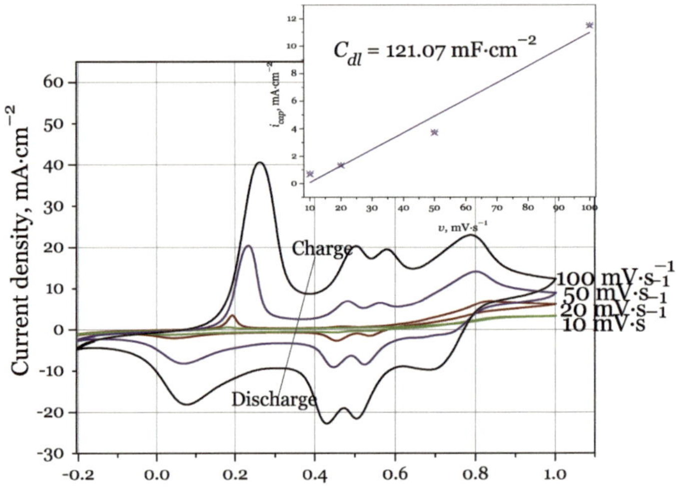

Figure 7.
Cyclic electrochemical spectra of pure PANI (0.1 M aniline +0.5 M H2SO4 + 3%wt. graphene related to aniline weight) at different scan rates: 10, 20, 50, and 100 mV·s⁻¹. In inset is shown the change of capacitance current density by the change of scan rate.

$$i_{cap.} = \frac{|i_{anodic}| + |i_{cathodic}|}{2}, \tag{2}$$

The change of the capacitance current density by the change of the scan rate is shown in the inset of **Figures 6** and 7. The double layer capacity of G/PANI nanocomposite was determined to be 121.07 $mF{\cdot}cm^{-2}$. This is about 70 times higher related to the double layer capacity of the pure PANI. Incorporation of only 3% graphene in the polymer matrix of PANI, the electrochemical characteristics of the nanocomposite can be remarkably improved. Except the sensing application, these nanocomposites have good potential in electrochemical and energy storage devices [46–48].

3. pH sensing

The PANI-based pH sensors, obtained with electropolymerization have been tested to follow the pH changes of the water, on a laboratory scale, displaying quite good potentiality (**Figure 8**) [32, 33]. Besides the pure PANI, also several nanocomposites MWCNTs/PANI and G/PANI containing 1, 3, 5, and 10 wt% of nanostructures were created. The devices require immersion in the water sample for 30 sec. For 1 minute a voltammogram was recorded. The sensing response was expressed by the anodic peak at positive current values. The applied potential range during the measurement was −0.8 to +0.8 V, performed with a scan rate of 0.05 $V{\cdot}s^{-1}$.

They are suitable for the analysis of samples with pH values in the range required by the target application (**Figure 9**) and they displayed the ability to detect pH in both laboratory (buffers and seawater simulating solutions) and real samples. Experimentally obtained PANI-based SPE-sensors, created in the FP7-COMMONSENS project, were compared with commercial PANI-based SPE [49].

Actually, it has been verified that the presence of many possible components in real seawater does not affect the efficiency of the device working and not disturbing the measurements. Even in the case of good agreement of the electrode response to the pH variation and the results of the laboratory pH meters, commonly used as a reference, it is important to improve the sensitivity of the systems in order to allow more precise results. This can be achieved by: (i) more controlled nanocomposite deposition on the electrode, (ii) obtaining a layer with fixed thickness, and (iii) constant and uniform covering degree of the sensing electrode surface.

The precision value obtained during field testing activities (**Figure 10**) showed that the SPE-based pH sensor needs to be optimized.

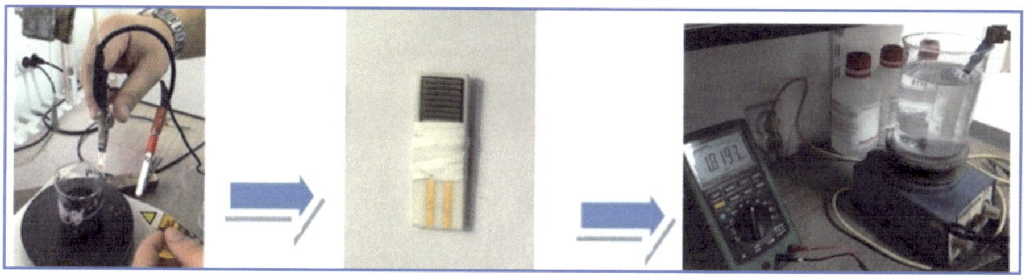

Figure 8.
PANI-based pH sensors, obtained with electro polymerization.

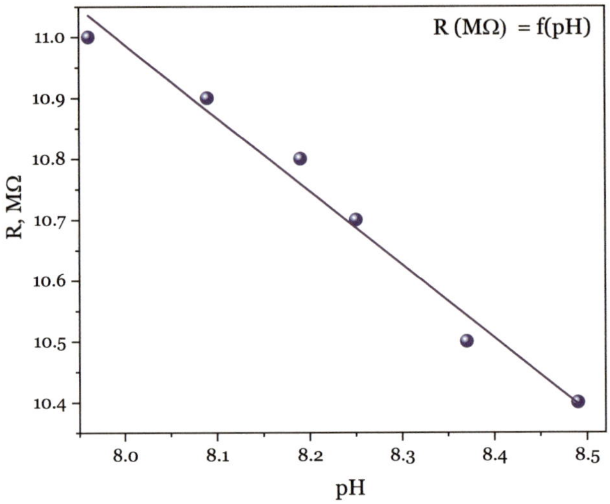

Figure 9.
Validation curve for pH range = 7.9 ÷8.5.

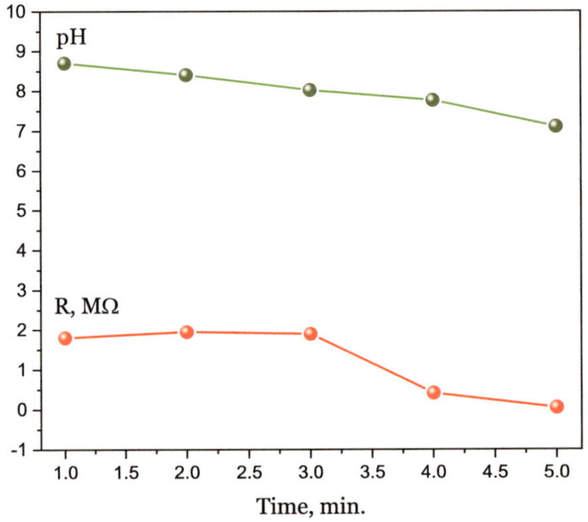

Figure 10.
Field testing at CNR in Oristano, September 2016, pH and electrical resistivity (MΩ] on abscissas and time [minutes] in ordinates.

The comparison of the measurements performed in the sample of the lab tank with the measurements performed directly in the lagoon has shown that both curves followed the same trend in the pH range of around 8 (**Figure 11**). Namely, the presented resistivity curve for 16 minutes at pH = 8.1 in the seawater was compared with the resistivity curve for 16 minutes in the lab tank when pH was changed in the range from 7.9 to 8.7 by adding NaOH drops in the water glass. It is evident that the resistivity values are in different ranges. For constant pH = 8.1 while for increasing pH from 7.9 to 8.7, resistivity in the lab tank was measured from 400 (pH = 7.9) to 500 kΩ (pH = 8.7).

Sample 3% wt MWCNT/PANI	R [MΩ]					
	1 min	2 min	3 min	5 min	10 min	12 min stabilized
First day	1.26	1.40	1.50	1.81	2.31	2.48
5 days	1.60	1.67	1.76	1.80	2.01	2.06
10 days	1.70	1.80	1.83	1.91	1.99	2.04
21 days	1.31	1.36	1.37	1.33	1.39	1.40

Table 2.
Stability of SPE-MWCNT nanosensors in seawater at pH = 8.4.

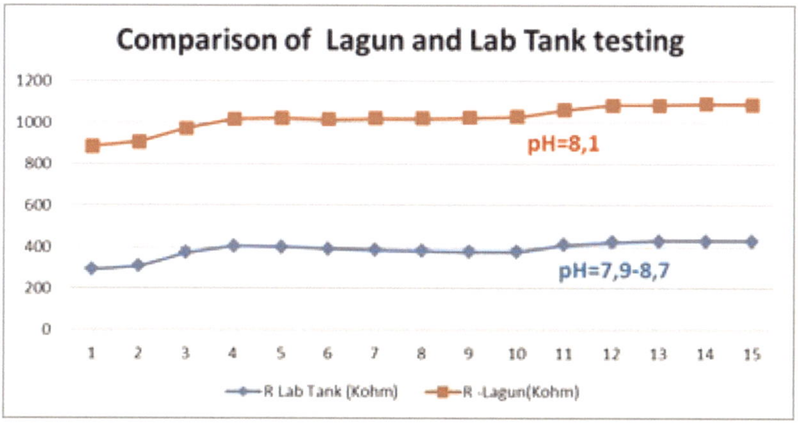

Figure 11.
Comparison of lagoon and lab tank testing by SPE of 3% G/PANI nanocomposite sensing element, with resistivity [kΩ] on abscissas and time [minutes] in ordinates.

Because by this technique a current is applied during the measurement, a more stable signal can be obtained. Once the applied current is optimized, the sensor needs to be tested using artificial samples. In the case of achieving good linear response, reproducibility, and precision values, the sensor can be tested using real seawater samples. The time needed to perform this action has been estimated to be of about 6–8 months. Also, with better resolution of the measurements, it can be possible to have better precision in the measured potential.

The stability of the pH sensor based on SPE of MWCNTs/PANI nanocomposites was tested in seawater at pH = 8.4. The obtained data are presented in **Table 2**.

3.1 Gas sensing

Industrialization, urbanization, and rapid technological modernization are three crucial factors contributing to over-exploitation of natural resources. With domination of the nonrenewable fossil fuels in the global energy supply, environmental pollution is worsening everywhere on the globe. In the past decade, atmospheric pollution has become a global phenomenon reaching life-threatening levels. Ammonia (NH_3) and nitrogen dioxide (NO2) are one of the most abundant inorganic toxic pollutants known for their worsening effect on the air quality, causing serious eye, skin, and mucous membrane respiratory irritation even when found in traces [50, 51]. Serving as a serious threat to human health, it is important to find an efficient and rapid solution for maintaining air quality in both indoor and outdoor environments. Researchers are constantly developing strategies and solutions for clean technology environmental applications to lower pollution damage. However, the first prerequisite procedure for environmental pollution treatment is monitoring. Therefore, it is imperative to develop novel sensors with outstanding characteristics such as higher sensitivity, selectivity, and accuracy, for effectively enriching pollutants in traces.

Even though many analytical instruments based on colorimetry, luminescence, or IR absorption are used for the concentration measurement of toxic gases, screen-printed electrochemical gas sensors are receiving constant attention because of their low-cost, simple design, and miniature size [52]. The screen-printing technique is one of the easiest and cheapest methods for chemiresistive sensor development which serves as an alternative to the traditional electrodes. The electrochemical principle requires a three-electrode system consisted of working, counter, and reference electrodes (**Figure 12**), where the sensing component or composite is deposited or printed onto the working electrode on various types of plastic or ceramic substrates. The principle is widely recognized and serves as one of the most promising approaches for sensor development because of its superior features—rapid in-situ analysis, high sensitivity and selectivity, portable miniaturized size, and low-cost [53]. Referred as economical electrochemical substrates, screen-printed electrodes are continually improving with respect to both their format and printing materials [52].

By now, there has been a large amount of research related to conducting polymer-based gas sensors, including our work [54] and other numerous reviews [2, 55–57]. Conductive polymers are explored as active substrates in gas sensors because of their broad properties in terms of conductivity, sensitivity, selectivity, and redox characteristics [17]. Among all of them, PANI, an intrinsically conducting polymer with excellent properties, has experience as a frequently investigated gas sensing component in chemiresitive sensors. PANI's nanodispersions have shown large potential for sensing applications, regarding the fact that they are inkjet and screen-printable, facilitating the conductive polymer pattern directly to the substrate [58]. Due to PANI's classification as a semiconductor, it is often used for accurate and rapid detection of reducing ammonia gas, a highly desirable task not only in environmental applications but also in the automotive and chemical industry and medical applications.

Chemoresistive sensors based on PANI have routinely been applied for ammonia detection, such as electrospun PANI fibers-based highly sensitive chemiresistive

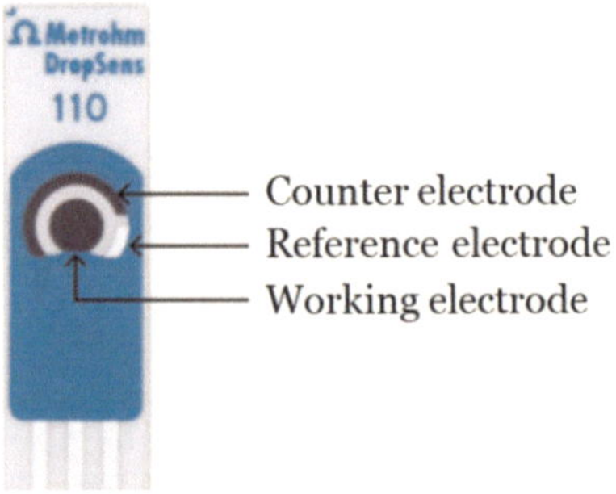

Figure 12.
Screen-printed electrode.

DOI: http://dx.doi.org/10.5772/intechopen.1002042

sensor described by Zhang et al. [59], MWCNT/PANI nanocomposite based SPE gas sensor proposed by Chepishevski et al. [60], inkjet-printed polyaniline nanoparticles based ammonia sensor described by Crowley et al. [61], modified gas sensor based on PANI nanofibers employing the techniques of electrical impedance spectroscopy with frequency response analysis and amperometry reported by Basak et al. [62] and many more. A highly sensitive and flexible ammonia gas sensor based on PANI film as an active sensing layer has been reported by Kumar et al. [63]. Their room temperature functioning sensor operated in the range of 5–1000 ppm, offered good reproducibility, long-term stability, and mechanical robustness, indicating the promising application of PANI films for portable on-site detection. Nanostructured PANI-based composites are suggested by Wojkiewicz and al. [64] for ppb range ammonia sensing. The author and his team studied three types of PANI composites with different morphologies: two core–shell systems with a poly(butyl acrylate) (PBuA) or poly (vinylidene fluoride) (PVDF) core and a PANI shell, and a composite based on PANI nanofibers embedded in a polyurethane (PU) matrix, and all of them showed high performances in terms of response time, reversibility and detection limit. The detection limit has been reported as below 100 ppb for films formed of the core–shell nanoparticles and below 20 ppb for the nanofiber-based composites. The team concluded that better sensitivity is achieved for the films made of the contacting core–shell nanoparticles. Talwar et al. [65] proposed a novel synthesis of ZnO-assisted PANI nanofibers and investigated their sensing response regarding ammonia gas. The fabricated sensor showed excellent selectivity and its sensing response has been proportional to the concentration of ammonia gas. Sutar et al. [66] prepared nanofibrous PANI films using an amino-silane self-assembled monolayer (SAM) employed as artificial seeds for the self-organization of PANI during polymerization. Chemiresistor sensors that have been developed using the fabricated nanofibrous PANI films as a sensitive layer, showed high sensitivity to very low concentrations (0.5 ppm).

One of the aims of this chapter is to show the results obtained from the electrochemical characterization of pristine PANI electrodes, as proposed sensors tested against ammonia (NH_3) vapors with different concentrations. The electrochemical characterization included resistance change monitoring of commercial PANI electrodes exposed to variable ammonia vapor concentrations of 3, 6.2, 12.5, and 25% (wt.). Commercial screen-printed PANI electrodes with 4 mm in diameter were ordered from Dropsens, Spain. In order to obtain better results, electrodes were exposed to external thermal excitation, measuring the sensing activity by heating and evaporating the ammonia solution around 50°C. The experimental setup scheme is shown in **Figure 13**.

The working mechanism is quite simple. The ammonia solution is evaporated in a closed circuit and ammonia gas is released. Because of the heating and humidity, the atmosphere gets more conductive, so it can absorb and more evenly distribute excess charges. Gas molecules travel and interact with the surface of the PANI electrode. When PANI is exposed to ammonia, its conductivity starts to change due to the deprotonation mechanism of amine groups in emeraldine salt converting it to emeraldine base following equation:

$$\mathrm{PANI} - \mathrm{H}^{+} + \mathrm{NH3} \leftrightharpoons \mathrm{PANI} + \mathrm{NH4}^{+}, \qquad (3)$$

proving excellent selectivity toward NH3 [67]. PANI's conductivity can be adjusted by changing its oxidation and protonation state and the material can be found in three key oxidative states: leucoemeraldine—fully reduced state, emeraldine—half-oxidized state, and pernigraniline—fully oxidized state.

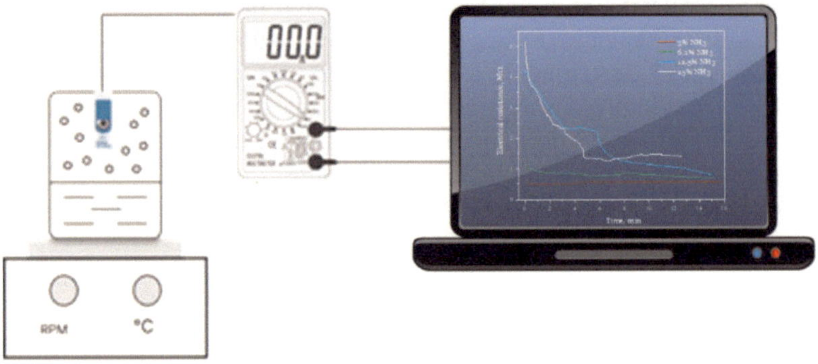

Figure 13.
Experimental setup for gas-sensing characterization.

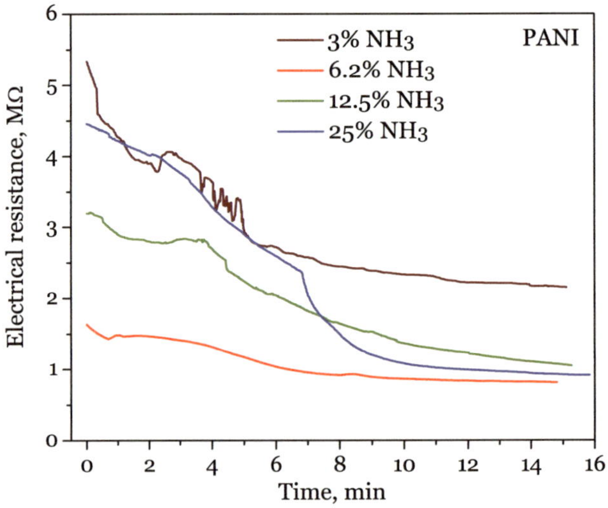

Figure 14.
Resistivity changes of commercial screen-printed PANI electrodes in various concentrations of NH3 ions.

The electrochemical characterization results obtained from the commercial PANI electrodes testing confirm electrochemical resistance decrease and conductivity increase, over time and at all concentrations. This effect is expected, as exposure to the ammonia vapors should indeed cause a resistance change which manifests as a decrease and conductivity increase.

Because of the conductive nature, PANI electrodes show high conductivity, as shown in **Figure 14**. Moreover, the tested electrodes show a non-linear response. The measured data suggests that it takes approximately 6 minutes to reach equilibrium, at all concentrations. This phenomenon happens because of the required time for the solution to create a humid environment. The sensor shows a stable and good response for all concentration levels. However, the highest conductivity is measured while testing the electrodes in the most concentrated ammonia solution i.e., 25%. The lowest response from the obtained measurement is shown for the ammonia solution with a concentration of 6.2%. After around 10 minutes, saturation is achieved at all concentrations and the curves show steadiness and linearity for the rest of the measurement.

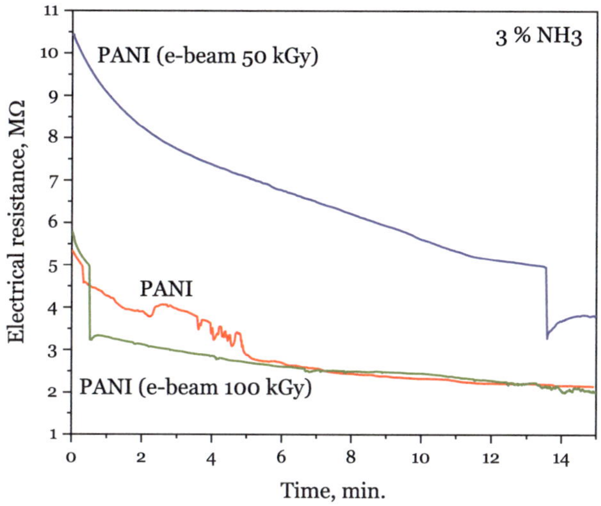

Figure 15.
Resistivity changes of irradiated SPE- PANI electrodes 3% of NH3 ions.

We tested the irradiation treatment of PANI electrodes by using 50 and 100 kGy e-beam irradiation. Comparison is given in **Figure 15**. Evidently, a higher effect was achieved with 50 kGy.

Looking through the literature data, this behavior agrees with the experimental results for pure PANI-based gas sensors [7, 68, 69]. PANI is found to exhibit detectable sensitivity expressed as increasing resistivity change for ammonia gas [18]. Improved response and sensitivity can be achieved by the addition of other conductive materials to the PANI matrix, which can act synergistically, such as carbon materials (CNT, graphene), metals (Ag, Au, Pt, Cu), inorganic nanoparticles (CeO_2, TiO_2, ZrO_2, Fe_2O_3, Fe_3O_4), chalcogenides (CdS, ZnS, CdSe), polymers (polyvinyl alcohol (PVA), polyvinyl acetate (PVAc) and polymethyl methacrylate (PMMA)), etc. [2, 56].

4. Conclusions

PANI and PANI-based nanocomposites with CNT and G were employed in the design of highly sensitive electrochemical sensors for the monitoring of the pH of sea and ocean waters and gases compounds. PANI and its nanocomposites possess excellent electrocatalytic properties for the modified sensors, such as enhanced detection sensitivity, electrocatalytic effects, high conductivity, and reduced fouling. These superior attributes endow CNTs, graphene, and PANI nanocomposites with great advantages for enhanced monitoring of pH and gas sensing applications. Carbon nanostructures (CNT, G) have shown that they may play a critical role in the future improvement of PANI-based sensor development of advanced points of other sensing applications.

Acknowledgements

The research reported in this chapter was part and financed by several projects: the FP7 COMMON SENSE (Fp7-614155) project, NANO IRA NET-MAK1003 from the

IAEA project, and the bilateral scientific project between the Faculty of Technology and Metallurgy—University Ss Cyril and Methodius in Skopje from the Republic of North Macedonia and Institute of Solid State Physics—University of Technology Graz from Austria.

Conflict of interest

The authors declare no conflict of interest.

Author details

Anita Grozdanov*, Perica Paunović, Iva Dimitrievska and Aleksandar Petrovski
Faculty of Technology and Metallurgy, University Ss Cyril and Methodius in Skopje, Skopje, North Macedonia

*Address all correspondence to: anita@tmf.ukim.edu.mk

References

[1] Abdulla S, Mathew TL, Pullithadathil B. Highly sensitive, room temperature gas sensor based on polyaniline-multiwalled carbon nanotubes (PANI/MWCNTs) nanocomposite for trace-level Ammonia detection. Sensors and Actuators B: Chemical. 2015;**221**: 1523-1534. DOI: 10.1016/j.snb.2015.08.002

[2] Beygisangchin M, Abdul Rashid S, Shafie S, Sadrolhosseini AR, Lim HN. Preparations, properties, and applications of polyaniline and polyaniline thin films-a review. Polymers. 2021;**13**:2003. DOI: 10.3390/polym13122003

[3] Li S, Lin P, Zhao L, Wang C, Liu D, Liu F, et al. The room temperature gas sensor based on polyaniline@flower-like WO3 nanocomposites and flexible PET substrate for NH3 detection. Sensors and Actuators B: Chemical. 2018;**259**: 505-513. DOI: 10.1016/j.snb.2017.11.081

[4] Das S, Sen B, Debnath N. Recent trends in nanomaterials applications in environmental monitoring and remediation. Environmental Science and Pollution Research. 2015;**22**: 18333-18344. DOI: 10.1007/s11356-015-5491-6

[5] Betty CA, Choudhury S, Arora S. Tin oxide–polyaniline heterostructure sensors for highly sensitive and selective detection of toxic gases at room temperature. Sensors and Actuators B: Chemical. 2015;**220**:288-294. DOI: 10.1016/j.snb.2015.05.074

[6] Sengupta PP, Barik S, Adhikari B. Polyaniline as a gas-sensor material. Materials and Manufacturing Processes. 2006;**21**:263-270. DOI: 10.1080/10426910500464602

[7] Fratoddi I, Venditti I, Cametti C, Russo MV. Chemiresistive polyaniline-based gas sensors: A mini review. Sensors and Actuators B: Chemical. 2015; **220**:534-548. DOI: 10.1016/j.snb.2015.05.107

[8] Doan DCT, Ramaneti R, Baggerman J, Van der Bent J, Marcelis ATM, Tong HD, et al. Carbon dioxide sensing with sulfonated polyaniline. Sensors and Actuators B: Chemical. 2012;**168**:123-130. DOI: 10.1016/j.snb.2012.03.065

[9] Barde RV. Preparation, characterization and CO2 gas sensitivity of polyaniline doped with sodium superoxide (NaO2). Materials Research Bulletin. 2016;**73**:70-76. DOI: 10.1016/j.materresbull.2015.08.026

[10] Parangusan H, Bhadra J, Ahmad Z, Mallick S, Touati F, Al-Thani N. Humidity sensor based on poly(lactic acid)/PANI–ZnO composite electrospun fibers. RSC Advances. 2021;**11**: 28735-28743. DOI: 10.1039/D1RA02842A

[11] He M, Xie L, Luo G, Li Z, Wright J, Zhu Z. Flexible fabric gas sensors based on PANI/WO3 p−n heterojunction for high performance NH3 detection at room temperature. Science China Materials. 2020;**63**:2028-2039. DOI: 10.1007/s40843-020-1364-4

[12] Shen S, Fan Z, Deng J, Guo X, Zhang L, Liu G, et al. An LC passive wireless gas sensor based on PANI/CNT composite. Sensors. 2018;**18**:3022. DOI: 10.3390/s18093022

[13] Parmar M, Balamurugan C, Lee DW. PANI and graphene/PANI nanocomposite films — Comparative toluene gas sensing behavior. Sensors.

2013;**13**:16611-16624. DOI: 10.3390/s131216611

[14] Gaikwad G, Patil P, Patil D, Naik J. Synthesis and evaluation of gas sensing properties of PANI based graphene oxide nanocomposites. Materials Science and Engineering: B. 2017;**218**:14-22. DOI: 10.1016/j.mseb.2017.01.008

[15] Zhu G, Zhang Q, Xie G, Su Y, Zhao K, Du H, et al. Gas sensors based on polyaniline/zinc oxide hybrid film for ammonia detection at room temperature. Chemical Physics Letters. 2016;**665**:147-152. DOI: 10.1016/j.cplett.2016.10.068

[16] Liu J, Cui N, Xu Q, Wang Z, Gu L, Dou W. High-performance PANI-based Ammonia gas sensor promoted by surface Nanostructuralization. Journal of Solid State Science and Technology. 2021;**10**:027007. DOI: 10.1149/2162-8777/abe3ce

[17] Macagnano A, Zampetti E, Pantalei S, De Cesare F, Bearzotti A, Persaud KC. Nanofibrous PANI-based conductive polymers for trace gas analysis. Thin Solid Films. 2011;**520**:978-985. DOI: 10.1016/j.tsf.2011.04.175

[18] Korent A, Žagar Soderžnik K, Šturm S, Žužek Rožman K, Redon N, Wojkiewicz JL, et al. Facile fabrication of an Ammonia-gas sensor using electrochemically synthesised polyaniline on commercial screen-printed three-electrode systems. Sensors. 2021;**21**:169. DOI: 10.3390/s21010169

[19] Zhu C, Xu Y, Zhou T, Liu L, Chen Q, Gao B, et al. Self-assembly polyaniline films for the high-performance ammonia gas sensor. Sensors and Actuators B: Chemical. 2022;**365**:131928. DOI: 10.1016/j.snb.2022.131928

[20] Jin Z, Su Y, Duan Y. An improved optical pH sensor based on polyaniline. Sensors and Actuators B: Chemical. 2000;**71**:118-122, ISSN 0925-4005. DOI: 10.1016/S0925-4005(00)00597-9

[21] Wei Y, Jaag G-W, Chan C-C, Hsueh KF, Hariharan R, Patel SA, et al. Polymerization of aniline and alkyl ring-substituted anilines in the presence of aromatic additives. The Journal of Physical Chemistry. 1990;**94**:7716-7721. DOI: 10.1021/j100382a073

[22] Gospodinova N, Terlemezyan L. Conducting polymers prepared by oxidative polymerization: Polyaniline. Progress in Polymer Science. 1998;**23**:1443-1484. DOI: 10.1016/S0079-6700(98)00008-2

[23] Bhadra S, Dipak Khastgir D, Singha NK, Lee JH. Progress in preparation, processing and applications of polyaniline. Progress in Polymer Science. 2009;**34**:783-810

[24] Neoh KG, Tan KL, Tan TC, Kang ET. Effects of protonic acids on polyaniline structure and characteristics. Journal of Macromolecular Science: Part A - Chemistry. 1990;**A27**:347-360. DOI: 10.1080/00222339009349558

[25] Yazdanpanah A, Ramedani A, Abrishamkar A, Milan PB, Moghadan ZS, Chauhan NPS, et al. Synthetic route of PANI (V): Electrochemical polymerization. In: Mozafari M, Chauhan NPS, editors. Fundamentals and Emerging Applications of Polyaniline. Amsterdam: Elsevier Inc.; 2019. pp. 105-119. DOI: 10.1016/B978-0-12-817915-4.00006-3

[26] Lamaoui A, García-Guzmán JJ, Palacios-Santander JM, Cubillana-Aguilera L. Synthesis techniques of molecularly imprinted polymer

composites. In: Sooraj MP, Nair AS, Mathew B, Thomas S, editors. Molecularly Imprinted Polymer Composites. Amsterdam: Elsevier Inc.; 2021. pp. 49-91. DOI: 10.1016/B978-0-12-819952-7.00002-0

[27] Vivekanandan J, Ponnusamy V, Mahudeswaran A, Vijayanand PS. Synthesis, characterization and conductivity study of polyanilinee prepared by chemical oxidative and electrochemical methods. Archives of Applied Science Research. 2011;**3**: 147-153 http://scholarsresearchlibrary.com/archive.html

[28] Gvozdenović MM, Jugović BZ, Stevanović JS, TLJ T, Grgur BN. Electrochemical polymerization of aniline. In: Schab-Balcerzak E, editor. Electropolymerization. Rijeka: InTech; 2011. pp. 77-96. DOI: 10.5772/28293

[29] Yoon SB, Yoon EH, Kim KB. Electrochemical properties of leucoemeraldine, emeraldine, and pernigraniline forms of polyaniline/multi-wall carbon nanotube nanocomposites for supercapacitor applications. Journal of Power Sources. 2011;**196**:10791-10797. DOI: 10.1016/j.jpowsour.2011.08.107

[30] Pharhad Hussain AM, Kumar A. Electrochemical synthesis and characterization of chloride doped polyaniline. Bulletin of Materials Science. 2003;**26**:329-334. DOI: 10.1007/BF02707455

[31] Sasaki K, Kaya M, Yano J, Kitani A, Kunai A. Growth mechanism in the electropolymerization of aniline and p-aminodiphenylamine. Journal of Electroanalytical Chemistry and Interfacial Electrochemistry. 1986;**215**: 401-407. DOI: 10.1016/0022-0728(86)87033-4

[32] Petrovski A, Cocca M, Paunović P, Avolio R, Errico ME, Barton J, et al. Synthesis and characterization of nanocomposites based on PANI and carbon nanostructures prepared by electropolymerization. Materials Chemistry and Physics. 2017;**185**:83-90. DOI: 10.1016/j.matchemphys.2016.10.008

[33] Petrovski A, Paunović P, Grozdanov A, Dimitrov AT, Mickova I, Gentile G, et al. Electrochemical polymerization and in situ characterization of PANI in presence of chemically modified graphene. Bulgarian Chemical Communications. 2020;**52**(E): 41-48

[34] LjD A, Plieth W, Koßmehl G. Electrochemical and Raman spectroscopic study of polyaniline; influence of the potential on the degradation of polyaniline. Journal of Solid State Electrochemistry. 1998;**2**: 355-361. DOI: 10.1007/s100080050112

[35] Mickova I, Prusi A, Grčev T. Arsov Lj. Electrochemical polymerization of aniline in presence of TiO_2 nanoparticles. Bulletin of the Chemists and Technologists of Macedonia. 2006;**25**:45-50. DOI: 10.20450/MJCCE.2006.278

[36] Taranu BO, Fagadar-Cosma E, Popa I, Plesu N, Taranu I. Adsorbed functionalized porphyrins on polyaniline modified platinum electrodes. Comparative electrochemical properties. Digest Journal of Nanomaterials and Biostructures. 2014;**9**:667-679

[37] Stilwell DE, Park SM. Electrochemistry of conductive polymers: II. Electrochemical studies on growth properties of polyaniline. Journal of the Electrochemical Society. 1988;**135**: 2254-2262. DOI: 10.1149/1.2096248

[38] Geniès EM, Lapkowski M, Penneau FJ. Cyclic voltammetry of

polyaniline: Interpretation of the middle peak. Journal of Electroanalytical Chemistry and Interfacial Electrochemistry. 1988;**249**:97-107. DOI: 10.1016/0022-0728(88)80351-6

[39] Pournaghi-Azar MH, Habibi B. Electropolymerization of aniline in acid media on the bare and chemically pre-treated aluminum electrodes: A comparative characterization of the polyaniline deposited electrodes. Electrochimica Acta. 2007;**52**:4222-4230. DOI: 10.1016/j.electacta.2006.11.050

[40] Gu J, Kan S, Shen Q, Kan J. Effects of Sulfanilic acid and Anthranilic acid on electrochemical stability of polyaniline. International Journal of Electrochemical Science. 2014;**9**:6858-6869

[41] Gajendran P, Saraswathi R. Polyaniline–carbon nanotube composites. Pure and Applied Chemistry. 2008;**80**:2377-2395. DOI: 10.1351/pac200880112377

[42] Malinauskas A, Malinauskienė J. Revising the kinetics of aniline electropolymerization under controlled potential conditions. Chemija. 2005;**16**:1-7

[43] Mu S, Kan J, Lu J, Zhuang L. Interconversion of polarons and bipolarons of polyaniline during the electrochemical polymerization of aniline. Journal of Electroanalytical Chemistry. 1998;**446**:107-112. DOI: 10.1016/S0022-0728(97)00529-9

[44] Yang H, Bard AJ. The application of fast scan cyclic voltammetry. Mechanistic study of the initial stage of electropolymerization of aniline in aqueous solutions. Journal of Electroanalytical Chemistry. 1992;**339**: 423-449. DOI: 10.1016/0022-0728(92)80466-H

[45] Da Silva ML, De Faria LA, Boodts JFC. Determination of the morphology factor of oxide layers. Electrochimica Acta. 2001;**47**:395-403. DOI: 10.1016/S0013-4686(01)00738-1

[46] Peng C, Zhang S, Jewell D, Chen GZ. Carbon nanotube and conducting polymer composites for supercapacitors. Progress in Natural Science. 2008;**18**: 777-788. DOI: 10.1016/j.pnsc.2008.03.002

[47] Zheng J, Ma X, He X, Gao M, Li G. Preparation, characterizations, and its potential applications of PANI/ graphene oxide nanocomposite. Procedia Engineering. 2012;**27**:1478-1487. DOI: 10.1016/J.PROENG.2011.12.611

[48] Kumar A, Kumar V, Awasthi K. Polyaniline–carbon nanotube composites: Preparation methods, properties, and applications. Polymer-Plastics Technology and Engineering. 2017;**57**:70-97. DOI: 10.1080/03602559.2017.1300817

[49] FP7 (OCEAN 2013.2) Project "Cost-effective sensors, inter-operable with internati nal existing ocean observing systems, to meet EU policies requirements" (Project reference 614155)

[50] Song X, Hu R, Xu S, Liu Z, Wang J, Shi Y, et al. Highly sensitive Ammonia gas detection at room temperature by Integratable silicon nanowire field-effect sensors. ACS Applied Materials & Interfaces. 2021;**13**:14377-14384. DOI: 10.1021/acsami.1c00585

[51] Zhang W, Cao S, Wu Z, Zhang M, Cao Y, Guo J, et al. High-performance gas sensor of polyaniline/carbon nanotube composites promoted by Interface engineering. Sensors. 2020;**20**: 149. DOI: 10.3390/s20010149

[52] Li M, Li YT, Li DW, Long YT. Recent developments and applications of

screen-printed electrodes in environmental assays—A review. Analytica Chimica Acta. 2021;**734**:31-44. DOI: 10.1016/j.aca.2012.05.018

[53] Domínguez Renedo O, Alonso-Lomillo MA, Arcos Martínez MJ. Recent developments in the field of screen-printed electrodes and their related applications. Talanta. 2007;**73**:202-219. DOI: 10.1016/j.talanta.2007.03.050

[54] Grozdanov A, Dimitrievska I, Paunović P, Petrovski A. Screen printed electrodes based on polymer/MWCNT and polymer/G nanocomposite for advanced gas sensing application. Material Science & Engineering International Journal. 2020;**4**:102-108. DOI: 10.15406/mseij.2020.04.00135

[55] Kwak D, Lei Y, Maric R. Ammonia gas sensors: A comprehensive review. Talanta. 2019;**204**:713-730. DOI: 10.1016/j.talanta.2019.06.034

[56] Sen T, Mishraa S, Shimpi NG. Synthesis and sensing applications of polyaniline nanocomposites: A review. RSC Advances. 2016;**6**:42196-42222. DOI: 10.1039/C6RA03049A

[57] Pandey S. Highly sensitive and selective chemiresistor gas/vapor sensors based on polyaniline nanocomposite: A comprehensive review. Journal of Science: Advanced Materials and Devices. 2016;**1**:431-453. DOI: 10.1016/j.jsamd.2016.10.005

[58] Crowley K, Smyth MR, Killard AJ, Morrin A. Printing polyaniline for sensor applications. Chemical Papers. 2013;**67**:771-780. DOI: 10.2478/s11696-012-0301-9

[59] Zhang Y, Kim JJ, Chen D, Tuller HL, Rutledge GC. Electrospun polyaniline Fibers as highly sensitive room temperature Chemiresistive sensors for Ammonia and nitrogen dioxide gases. Advanced Functional Materials. 2014;**24**:4005-4014. DOI: 10.1002/adfm.201400185

[60] Chepishevski G, Petrovski A, Grozdanov A, Paunović P, Dimitrov A, Gentile G, et al. MWCNT/PANI screen printed electrodes for gas sensors. In: Petkov P, Tsiulyanu D, Popov C, Kulisch W, editors. Advanced Nanotechnologies for Detection and Defence against CBRN Agents. Dordrecht: Springer; 2018. pp. 389-396. DOI: 10.1007/978-94-024-1298-7_38

[61] Crowley K, Morrin A, Hernandez A, O'Malley E, Whitten PG, Wallace GG, et al. Fabrication of an ammonia gas sensor using inkjet-printed polyaniline nanoparticles. Talanta. 2008;**77**:710-717. DOI: 10.1016/j.talanta.2008.07.022

[62] Basak SP, Kanjilal B, Sarkar P, Turner APF. Application of electrical impedance spectroscopy and amperometry in polyaniline modified ammonia gas sensor. Synthetic Metals. 2013;**175**:127-133. DOI: 10.1016/j.synthmet.2013.05.011

[63] Kumar L, Rawal I, Kaur A, Annapoorni S. Flexible room temperature ammonia sensor based on polyaniline. Sensors and Actuators B: Chemical. 2017;**240**:408-416. DOI: 10.1016/j.snb.2016.08.173

[64] Wojkiewicz JL, Bliznyuk VN, Carquigny S, Elkamchi N, Redon N, Lasri T, et al. Nanostructured polyaniline-based composites for ppb range ammonia sensing. Sensors and Actuators B: Chemical. 2011;**160**:1394-1403. DOI: 10.1016/j.snb.2011.09.084

[65] Talwar V, Singh O, Singh RC. ZnO assisted polyaniline nanofibers and its application as ammonia gas sensor.

Sensors and Actuators B: Chemical. 2014;**191**:276-282. DOI: 10.1016/j.snb.2013.09.106

[66] Sutar DS, Padma N, Aswal DK, Deshpande SK, Gupta SK, Yakhmi JV. Preparation of nanofibrous polyaniline films and their application as ammonia gas sensor. Sensors and Actuators B: Chemical. 2007;**128**:286-292. DOI: 10.1016/j.snb.2007.06.015

[67] Kukla AL, Shirshov YM, Piletsky SA. Ammonia sensors based on sensitive polyaniline films. Sensors and Actuators B: Chemical. 1996;**37**:135-140. DOI: 10.1016/S0925-4005(97)80128-1

[68] Pang Z, Yildirim E, Pasquinelli MA, Wei Q. Ammonia sensing performance of polyaniline-coated polyamide 6 nanofibers. ACS Omega. 2021;**6**: 8950-8957. DOI: 10.1021/acsomega.0c06272

[69] Hirata M, Sun L. Characteristics of an organic semiconductor polyaniline film as a sensor for NH3 gas. Sensors and Actuators A: Physical. 1994;**40**:159-163. DOI: 10.1016/0924-4247(94)85024-0

Chapter 6

Synthesis, Structural Study and Various Applications of Polyaniline and its Nanocomposites

Gobind Mandal, Jayanta Bauri, Debashish Nayak, Sanjeev Kumar, Sarfaraz Ansari and Ram Bilash Choudhary

Abstract

The long lasting intrinsic conducting polymers (ICPs) including polyaniline (PANI), polypyrole (PPy), Polyindole (PIn), Poly (methyl methacrylate) (PMMA), Polythiophene (PT), poly (3,4-ethylene dioxythiophene) (PEDOT) have been recognized for their significant benefits in optoelectronic devices. In the last few decades, polyaniline has gained recognition over metals, owing its low cost, flexibility, and high conductivity, as well as the ease with which it may be produced using chemical or electrochemical processes. Due to its high electrical conductivity, light weight, ease of fabrication, and excellent environmental stability, PANI has an extensive range of applications, including batteries, sensors, supercapacitors, waste water treatment and organic electronic devices. It also has the potential for chemical and electrochemical synthesis. Polyaniline has promising potential in many optoelectronic applications as well as in supercapacitors. In this chapter, the basic historical background, different synthesis mechanism about conducting polymer polyaniline is discussed in details. Polyaniline has great potential application such as in sensors, supercapacitor and optoelectronic devices etc. due to its ability of ease of synthesis by various methods. Polyaniline based nanocomposites with different metals, metal oxide, metal sulfides, and carbon nanomaterials, graphene, carbon nanotubes (CNTs) etc. are described in this section in detail.

Keywords: polyaniline, protonation, chemical oxidative polymerization, OLEDs, supercapacitors

1. Introduction

Since ancient times researchers have been trying to develop new smart materials which have unique amalgamations of properties. In modern times, conducting polymer-based nanocomposites and nano blending technologies were extensively utilized in various fields such as supercapacitors, solar cells, sensors, and different types of electronic device applications. Conducting polymer becomes promising candidates due to their unique properties such as transparency, lightweight, low cost, flexibility, optical, electrical, and dielectric over the inorganic semiconductors [1]. Uses of inorganic semiconductors are replaced by conducting polymers all over the technologies. Intrinsic conducting polymers are divided into three types according to the conduction mechanism: Π conjugated, redox, and ionic conducting polymers [2]. In Π conjugated systems, the electron is carriers to conduct electricity through the backbone skeleton of the polymer. Thus, these electrons are called delocalization of electrons. The redox polymers contain immobilized redox or electroactive centers. These electroactive centers are not connected to each other. Charge (electrons) are transferred in this system via a hopping mechanism from one active center to the other. To increase the conductivity, systems need to have large numbers of redox-active centers. In the case of ionic polymers contain ions. This system conducts electricity due to the movements of ions in the polymer chain [3]. The conducting polymers' electrical conductivity, mechanical stability, and optical properties were enhanced multiple times by doping or mixing inorganic semiconductors as filler elements. Recently, inorganic-based polymeric nanocomposites have been extensively used in various fields [2, 3].

Hideki Shirakawa, Alan Heeger, and Alan Mac Diarmid discovered the simplest conducting polymer in 1977 and got the Nobel Prize in 2000 [4]. After that, a new novel conducting polymer, polymerization mechanism, and electron transport mechanism gradually developed in front of researchers. Polyacetylene was the first synthesized conducting polymer; after that, several conducting polymers were developed, such as polythiophene, PEDOT: PSS, polypyrrole, polyaniline, and polycarbazole. Among these conducting polymers, polyaniline is the most reported conducting polymer in electronic devices like sensors, solar cells, supercapacitors, anticorrosion materials, and optoelectronic device applications [5, 6]. Henry Letheby first developed aniline black or PANI by using HNO3 as an oxidant agent via the oxidative chemical polymerization method, which is insulating in nature. In 1997 Mac Diarmid introduced another derivative (green emeraldine salt), the aniline family, which conducts electricity. He introduced three different derivative aniline families one is colorless reduced leucoemeraldine base (LEB), the second is blue/green half oxidized emeraldine base/salt (EB), and the third one is blue/violet fully oxidized pernigraniline base (PAB). Emeraldine salt is extensively used out of these three derivatives because it has outstanding optical, electrical, and anticorrosive properties [7]. Optical band gap energy (Eg = 1.20 eV–3.2 eV) and electrical conductivity (σele = 10–8 s/cm–10 s/cm) depend on the synthesis method. The refractive index of polyaniline is 1.31–1.36. The polyaniline exhibited a different type of nanostructure (such as tower-like, nanoflower-like, leaf-like, rod-like, and sea urchin-like morphologies) depending on the synthesis techniques [8]. Different type of microstructure has been shown in **Figure 1**.

In this book chapter, we have discussed the basic historical background of conducting polymers, and the main focus of our work is details about conducting polymer polyaniline. Polyaniline based nanocomposites with different metals, metal oxide, metal sulfides, and carbon nanomaterials, graphene, carbon nanotubes (CNTs), reduced graphene oxide (rGO) etc. are described in this section. Different types of synthesis methods and polymerization mechanisms were broadly discussed. In the last section, we discussed the application of polyaniline, specifically in supercapacitor and optoelectronic device applications.

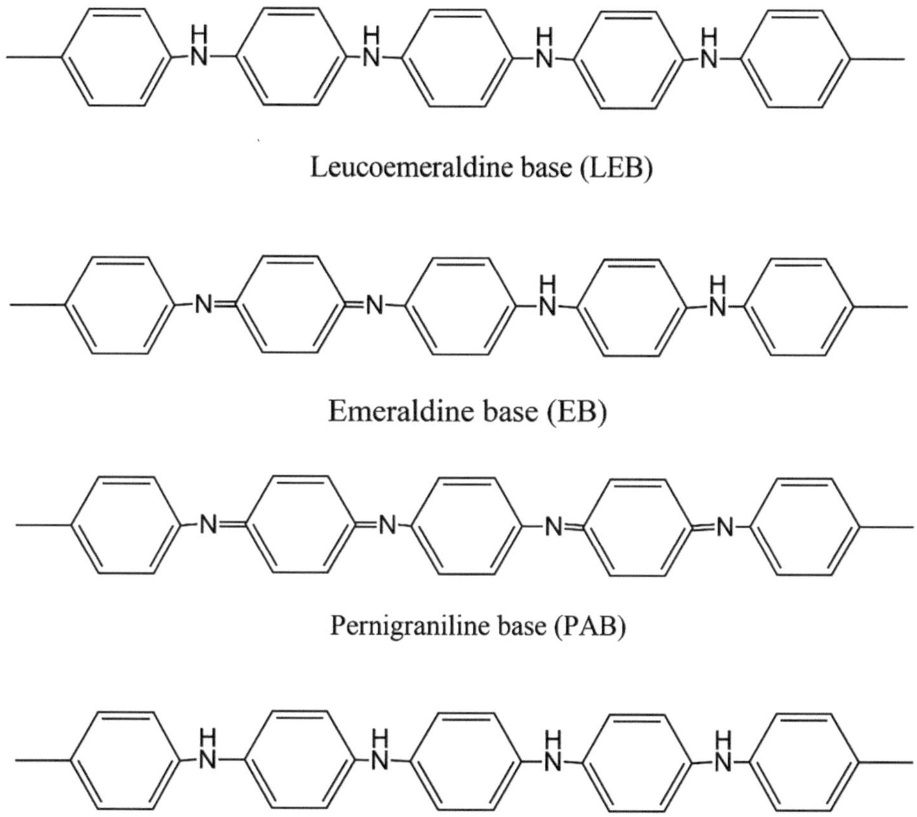

Figure 1.
Different derivatives of PANI.

2. Synthesis of polyaniline (PANI)

PANI is one of the promising and electrically conductive polymers owing its conversion ability between base and salt form. Access of oxidation and reduction and ease of synthesis made polyaniline more popular that attracts great deal of research. The most widely used synthesis method for preparing polyaniline is chemically oxidative polymerization technique using aniline monomer in acidic medium. The various method can be used for the synthesis of polyaniline are following:

1. Chemically oxidative polymerization

2. Electrochemical polymerization

3. Vapor-phase polymerization

4. Photochemically initiated polymerization

2.1 Chemically oxidative polymerization (COP)

For preparing monomers to polymer such as aniline to polyaniline, pyrrole to polypyrrole etc. chemical oxidative polymerization plays most prominent role. It is a cheap method for preparing large quantities of polymers with less time. In this method, chemical oxidizing agent initiates the polymerization process. The monomer compounds exhibit high electron donating properties. Chemical oxidative polymerization has been done with the help of oxidizing agents such as ammonium persulfate (APS), potassium dichromate, hydrogen peroxide etc. Ammonium persulfate $(NH_4)_2S_2O_8$ is a strong oxidizing agent which has been used to prepare polyaniline (PANI) from aniline monomer in our research work as it highly soluble in water. APS helps in generating radical cation sites in monomer and initiating the polymerization process. The prepared monomer solution is kept within aqueous acidic medium at a low temperature. This reaction is later allowed on a constant stirring for 5–8 hours in order to obtain the polymer precipitate (**Figure 2**) [9, 10].

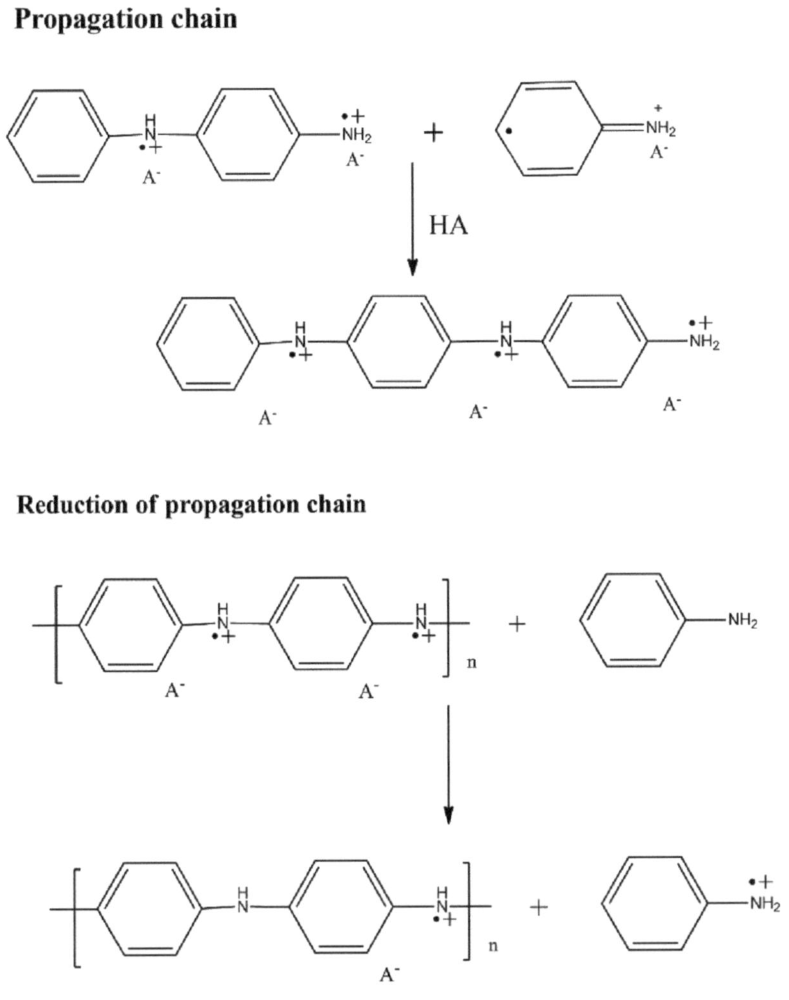

Figure 2.
Chemically oxidative polymerization of PANI.

2.2 Electrochemical polymerization (ECP)

For the preparation of thin film polyaniline with a huge surface area electrochemical process plays a vital role. Electrochemical method is quite similar to electrodeposition method used for metals. Electrochemical method provides homogenous polymer deposition over the electrode which contains a power source, an electrode and an electrolyte solution. Electrochemical polymerization of aniline is carried out in a strong acidic electrolyte such as acetonitrile which helps in the formation of anilinium radical cation by aniline oxidation on the electrode (**Figure 3**) [11].

2.3 Vapor phase polymerization (VPP)

Vapor phase polymerization is used to prepare polyaniline from aniline monomer by introducing to an oxidant coated substrate in vapor form. At the oxidant vapor interface, the polymerization takes place. Vapor phase polymerization (VPP) can be either chemical vapor phase polymerization (CVPP) or electrochemical vapor phase polymerization EVPP). CVPP is a solvent free process use to get highly uniform

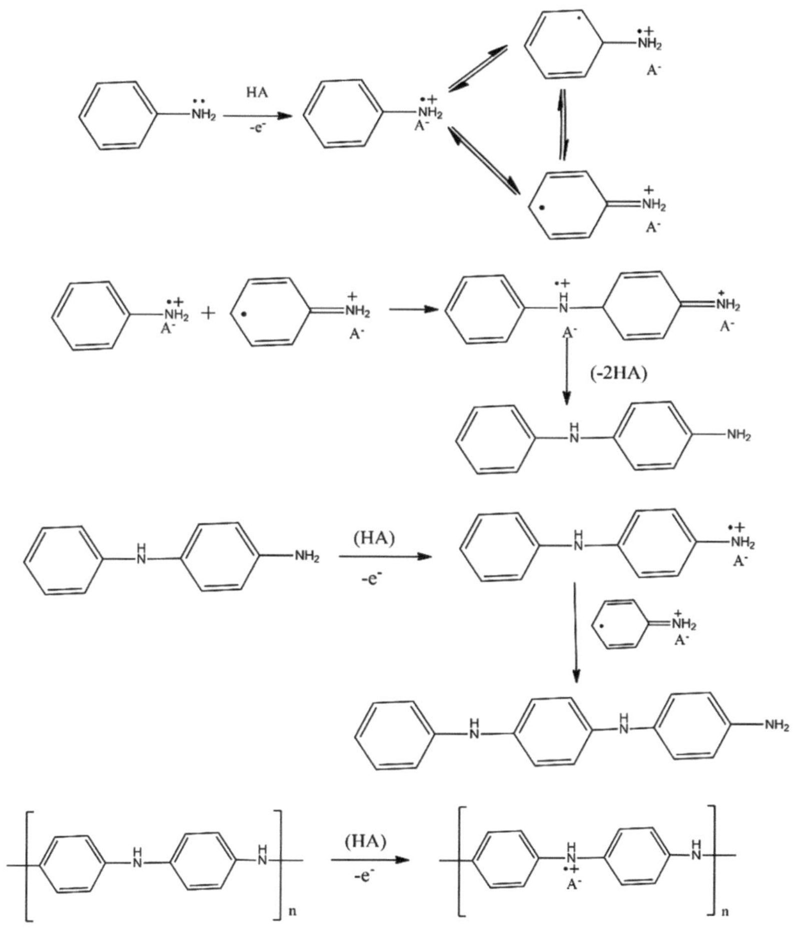

Figure 3.
Electrochemical polymerization of PANI.

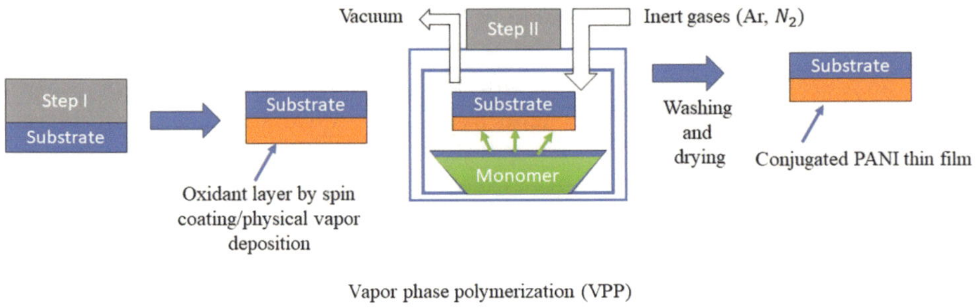

Figure 4.
Vapor phase polymerization method.

conductive polyaniline which allows formation of PANI layers of any thickness on an insulating substrate. In VPP monomers are applied as vapor rather than solution or liquid which restricts the particle agglomeration. Intrinsic conjugated polymers formed via VPP have high electrical conductivity and they are agglomeration free [12, 13]. VPP provides polymers of very high purity with excellent conductivity and scope of synthesis at the nanoscale range (**Figure 4**) [14].

2.4 Photochemically initiated polymerization

In this method light (photon) is used to initiate free radical polymerization. This technique consumes less energy and higher productivity at lower reaction temperature compared to other conventional polymerization techniques. In this method, the photochemical initiation is achieved by subjecting suitable photo initiators to ultra-violet radiations. The formation of PANI using photopolymerization requires an external source of gamma rays, microwaves, UV rays and X-rays. PANI can be synthesized using bilayer films containing $[Ru(bipy)_3]^{2+}$ as a primer and methylviologen (MV^{2+}) as an oxidizer polyaniline through irradiated with visible electromagnetic light. The process of electron transfer between $[Ru(bipy)_3]^{2+}$ and (MV^{2+}) is responsible for the oxidation of aniline and formation of PANI [15].

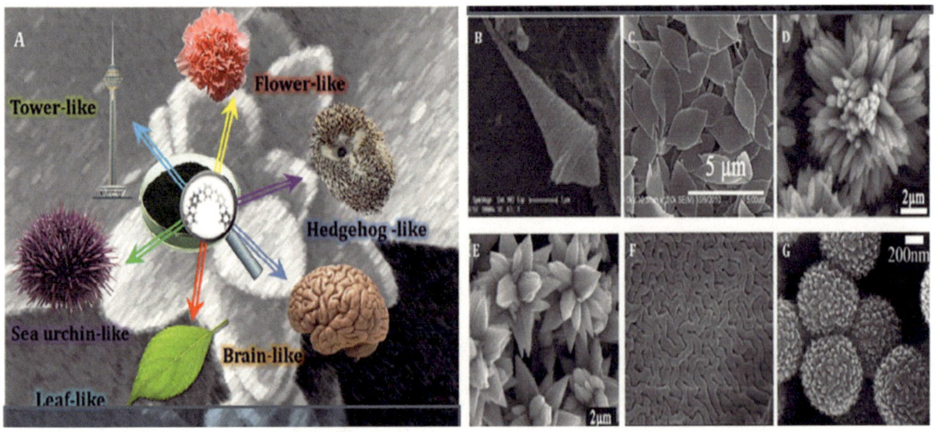

Figure 5.
(a) Various nanostructures/microstructures of PANI; SEM image of (b) tower/spindle-like PANI, (c) leaf-like PANI, (d) hedgehoglike PANI, (e) flower-like PANI, (f) brain-like PANI, and (g) sea urchin-like PANI [9].

The photochemical polymerization is similar to the free radical polymerization but initiation via light.

The different morphologies of PANI can be obtained using different synthesis method as shown in **Figure 5.**

3. Different methods for synthesis of nanocomposites

Synthesis of nanocomposites involves the combination of two or more materials at the nanoscale to create new materials with enhanced properties. There are various methods for synthesizing nanocomposites, and here are some commonly used techniques:

a. In-situ polymerization: This method involves the formation of nanoparticles within a polymer matrix during the polymerization process. Monomers and nanoparticles are combined, and polymerization occurs simultaneously, resulting in a nanocomposite material.

b. Solution mixing: In this method, nanoparticles and a polymer are separately dissolved in a common solvent and then mixed together. The solvent is subsequently removed, resulting in the formation of a nanocomposite.

c. Melt mixing: This technique involves the direct blending of nanoparticles and a polymer in the molten state. The mixture is then cooled and solidified, leading to the formation of a nanocomposite material.

d. Electrospinning: Electrospinning is a versatile method used to fabricate nanofibers. It involves the application of an electric field to a polymer solution or melt, resulting in the formation of ultrafine fibers. Nanoparticles can be incorporated into the polymer solution or melt prior to electrospinning to create nanocomposite fibers.

e. Layer-by-layer assembly: This technique involves the sequential deposition of alternating layers of nanoparticles and polymers. Electrostatic interactions or other forces are utilized to form the multilayered structure, resulting in the creation of nanocomposite films or coatings.

f. Sol-gel method: The sol-gel process involves the transformation of a sol (a dispersion of nanoparticles in a liquid) into a gel (a three-dimensional network of interconnected nanoparticles) by hydrolysis and condensation reactions. By controlling the reaction conditions, nanocomposite materials can be obtained.

g. Chemical vapor deposition (CVD): CVD is a technique used to deposit thin films of materials onto a substrate. In the case of nanocomposites, nanoparticles are introduced into the gas phase and allowed to react and deposit onto the substrate surface, resulting in the formation of a nanocomposite film.

h. Template synthesis: This method involves using a template or scaffold to guide the growth of nanoparticles or polymers. The template can be a porous material or a sacrificial structure that is subsequently removed, leaving behind a nanocomposite material with a specific structure.

These are just a few examples of the methods used for synthesizing nanocomposites. The choice of method depends on factors such as the desired properties of the nanocomposite, the nature of the materials involved, and the intended application.

4. Polyaniline-based nanocomposites

Polyaniline (PANI) is a conducting polymer with unique electrical, optical, and chemical properties. When combined with nanomaterials, such as nanoparticles or nanofibers, PANI-based nanocomposites exhibit enhanced properties and a wide range of potential applications. Polyaniline-based nanocomposites can be synthesized through various methods, including in situ polymerization, solution blending, electrochemical deposition, and template-assisted synthesis. The choice of synthesis method depends on the desired properties and application requirements. PANI can be combined with different types of nanomaterials, including metal nanoparticles (e.g., gold, silver), metal oxides (e.g., titanium dioxide, zinc oxide), carbon-based materials (e.g., carbon nanotubes, graphene), and organic/inorganic nanofibers. These nanomaterials serve as fillers or reinforcements, imparting unique properties to the nanocomposite. Enhanced electrical conductivity: PANI itself is a conductive polymer, but the addition of nanomaterials further enhances the electrical conductivity of the nanocomposite. This property makes PANI-based nanocomposites suitable for applications such as sensors, actuators, electrostatic discharge materials, and energy storage devices. Improved mechanical properties: The incorporation of nanomaterials in PANI matrices improves the mechanical strength and toughness of the nanocomposite. Nanofillers act as reinforcing agents, reducing the brittleness of PANI and enhancing its resistance to mechanical stress. Enhanced thermal stability: PANI exhibits limited thermal stability at high temperatures. However, the addition of nanomaterials can improve the thermal stability of PANI-based nanocomposites, allowing their use in applications that require higher operating temperatures. Tailored optical properties: PANI-based nanocomposites can exhibit tunable optical properties, including changes in color, absorbance, and emission, depending on the nanomaterials incorporated. These properties make them suitable for applications such as optoelectronics, light-emitting devices, and sensors. Polyaniline-based nanocomposites find applications in various fields, including energy storage and conversion (batteries, supercapacitors, solar cells), electronics (printed circuit boards, conductive coatings), sensors and actuators, corrosion protection, electromagnetic shielding, and biomedical applications (drug delivery, tissue engineering). Despite their promising properties, the development of PANI-based nanocomposites faces challenges related to achieving uniform dispersion of nanofillers, controlling the size and morphology of nanoparticles, and maintaining stability over time. Additionally, the scale-up of synthesis methods and the cost-effectiveness of production are areas that require further research and development. Polyaniline-based nanocomposites offer exciting opportunities for advancing materials science and developing innovative technologies. Continued research and exploration of these materials will likely lead to further improvements and new applications in the future. In past decades, the incorporation of inorganic filler into organic conducting polymer forming a composites or nanocomposites matrix attracted huge attention. The combination of inorganic provides better performance, stability and electrical conductivity to polymer matrix and has application in sensors, actuators, OLEDs, solar cells, supercapacitors and so on. Many publications have been reported for the synthesis of PANI with metal oxides, metal, sulfides, graphene and other inorganic materials.

4.1 PANI-based binary/ternary nanocomposites

Polyaniline nanocomposites can be considered of as a material composed of a PANI matrix and one component, including semiconductors, metal nanoparticles, organic compounds, inorganic compounds, biological and natural compounds, to modify the surface morphologies and enhanced the polymer stability, optical and electrical properties. Similarly, in case of ternary nanocomposites surface interaction enhanced modify the electronic structure [16]. In **Figure 6** shows the PANI based nanocomposites revealed from the literature survey.

PANI-based binary nanocomposites: Researchers synthesized and developed new electrode materials by refining different parameters in response to the rising demand for the development of high-performance supercapacitors.

4.2 Applications of polyaniline

Polyaniline exhibited in three different oxidation states such as Leucomeraldine (LM), pernigraline (PN) and emeraldine (EM). Emeraldine salt is a conducting nature polymer. Due to different surface morphology (such as nanofiber, nanoflowers, nano- leaf) of polyaniline shows distinct physiochemical properties. Large specific surface area, easily dissolved in alcohol, ketone and other organic solvent of polyaniline can improve the reactivity in the application of supercapacitor, sensor and water pollutant free applications. Due to semiconducting nature, excellent optical, electrical properties and higher thermal stability polyaniline has extensive application in the field of optoelectronic device (like organic light emitting diode (OLED), organic solar cell). In **Figure 6** shows the various applications of PANI based nanocomposites. Engineering is the current focus of medicine, and advancements in this field need for new intellectual technologies. Devices that compensate for nerve weakening and advance neuroscience are needed by neuroscientists. Biocompatibility conductive scaffolding has high bio-counterfeit qualities, and it has been used to treat organ problems. Moreover, PANI applications in delivery systems have drawn a lot of interest; as a consequence, novel delivery structures, such electro-drug delivery systems, are being investigated. PANI has been successfully used as an anti-corrosion barrier with positive outcomes [17]. **Figure 7** showed the application of the PANI in different field.

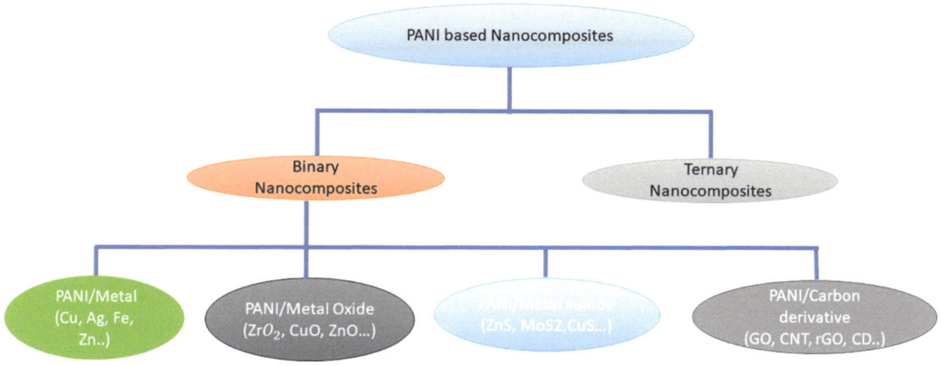

Figure 6.
PANI based nanocomposites.

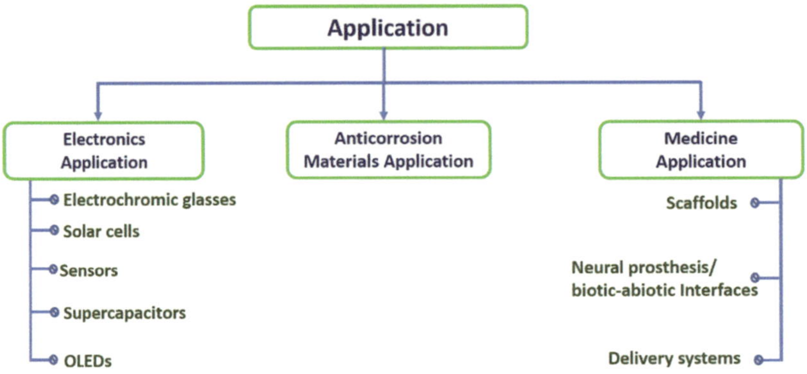

Figure 7.
Various applications of PANI based nanocomposites.

4.2.1 Supercapacitor applications

Supercapacitor is one of the highly anticipated energy storage technologies having numerous advantages over its most common counterpart i.e., rechargeable battery. It possessed quick energy supply, long life-span and environment compatibility. These properties suggest this technology as the energy storage for modern era [18, 19].

Conventional carbon material based supercapcitor (electric double layer capacitor) suffer from low specific capacitance and energy density [20, 21]. Therefore, recently researchers have turn towards conducting polymer based supercapcitor (pseudocapacitor) which provides enhanced specific capacitance due to presence of numerous redox active sites. Zhang et al. synthesized PANI via chemical oxidative polymerization. The resultant polymeric electrode exhibited specific capacitance of 330 Fg^{-1} with capacitance retention of 75% [22]. Similarly, Khdary et al. synthesized mesoporous PANI which passed much improved specific capacitance (532 Fg^{-1}) and cyclic stability (85% after 1000 cycles) [23]. The mesoporous nature of the material proved to be beneficial for charge mobility and increase its charge accumulation ability. However, PANI in its pristine form is not usually recommended for supercapacitor due to lack of stability [24]. The stability material can be improved by introducing different filler materials such as carbon materials, metal oxides, metal sulfides, metal organic frameworks (MOFs) etc. For instant, Shafi et al. synthesized ternary composite of $LaMnO_3$/RGO/PANI having excellent cyclic stability and energy density [25]. The polymeric material possessed specific capacitance of 802 Fg^{-1} at current density of 1 Ag^{-1}. Further the asymmetric supercapcitor (ASC) employing $LaMnO_3$/RGO/PANI as positive electrode and RGO as negative electrode showed highly improved cyclic behavior with capacitance retention of 117% even after 100 k cycles as represented in **Figure 8**. **Table 1** summarized electrochemical properties of some important PANI based electrode materials for supercapacitor applications.

4.2.2 Organic solar cell applications

Inexpensive, high-performance film solar cells based on an electrolyte and a semiconductor are known as dye-sensitive solar cells (DSSCs) [30, 31]. They typically include a counter electrode, which is often composed of platinum, a redox electrode, and a color-sensitive titanium dioxide electrode. Pt, which is replaced with

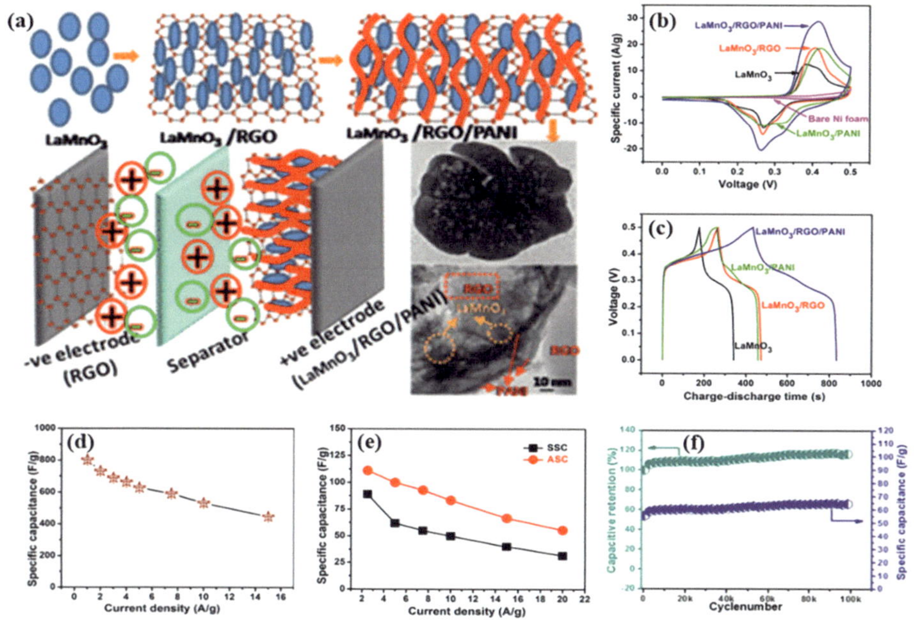

Figure 8.
(a) Schematic representation of $LaMnO_3$/RGO/PANI composite and its asymmetric supercapacitor; (b) CV curves for different electrode materials; (c) GCD curves for different electrode materials; (d) variation of specific capacitance of $LaMnO_3$/RGO/PANI electrode with current density; (e) variation of specific capacitance of symmetric and asymmetric supercapacitor devices with current density; and (f) cyclic stability of ASC up to 100 k cycles. Adapted with permission from [25]. Copyright 2018, American Chemical Society.

Sl. No.	Nanocomposite	Current density (Ag^{-1})	Specific capacitance (Fg^{-1})	Capacity retention	Ref.
1.	PANI	1	330	75% after 1000 cycles	[22]
2.	Mesoporous PANI	1.5	532	85% after 1000 cycles	[23]
3.	PANI/GO/Cu	1	558	90% after 1000 cycles	[26]
4.	PANI/RGO/$LaMnO_3$	1	802	117% after 100 k cycles	[25]
5.	PANI/Graphene/MnO_2	—	395	92% after 1200 cycles	[27]
6.	PANI-Ag/ZnO	0.8	635	96%	[28]
7.	PANI/graphene/ manganese ferrite	0.2	454	76.4% after 5000 cycles	[29]

Table 1.
Few electrochemical properties of PANI based nanocomposites as electrode materials for supercapacitor applications.

carbon-based materials, is the component in DSSCs that costs the highest. Because of its simple synthesis, cheap cost, and high conductivity, PANI is used in DSSC. Due to complicated electrocatalytic rooting activity at I3- and a lower charge transfer ratio of oxidation reactions, microporous PANI electrodes seem better than Pt electrodes [32]. The catalytic activity advances, the absorbency of PANI rises at a certain surface area, and the effectiveness of trapping liquid electrolytes for DSSC rises. In order to produce counter electrodes, PANI is electrolyzed in FTO glass by a number of counteractions, such as SO42-, BF4-, CL-, ClO4-, and p-toluene sulfonate [TsO-]. As a result, PANI-SO4 offers the greatest amount of porous medium while requiring the least amount of charge transfer resistance and the greatest amount of reduction current [33]. The electrical area of PANI and total conductivity are increased by PANI polymerization on the graphene surface [34]. Hence, an anti-electrode in DSSC or other electrocatalytic activity is improved by this strategy.

The absorbance range of the active layer material of organic solar cell should be in visible range of electromagnetic spectrum for the better efficiency of the device. The device with PSSA-g-PANI exhibits about 4% PCE, which is 20% higher than that of the device with PEDOT: PSS due to unique high transparency in the UV-vis region (especially 450–650 nm) and high conductivity of PSSA-g-PANI, when the blend of P3HT and PCBM is used as the active layer of PSCs as shown in **Figure 9**.

4.2.3 Organic light emitting diode (OLED) application

OLED is a device that produces light as a result of the induction of an electric current or electric field are known as electroluminescence devices. When a suitable voltage is applied, they can serve as a source for organic light-emitting diodes (OLEDs) with a p-n connection diode. As a result, photon energy can be discharged by combining electrons with the device's electron holes [35]. With the semiconductor energy bandgap, a vivid color will be produced. The organic LED (OLED) is a perforated injection film made of conductive polymer. OLED performance is improved

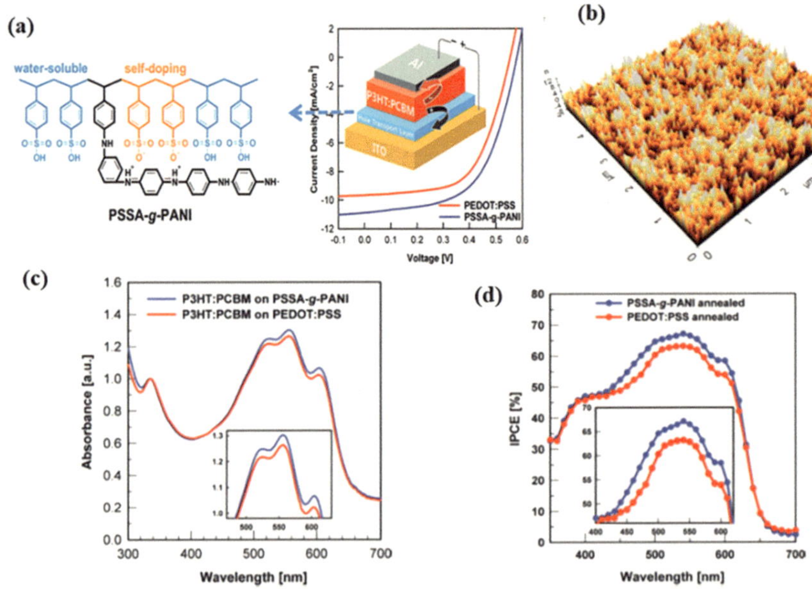

Figure 9.
(a) Device structure of solar cell. (b) AFM of the PSSA-g-PANI layer, (c) absorbance spectra of the device and (d) efficiency of the device.

by using PANI-poly (styrene sulfonate; PPS). Comparing PANI-PPS to commercial PEDOTPPS, more complex performance is obtained, and hole injection increases. PANI-PPS, in contrast, performs with the highest efficiency and lowest voltage given the high conductivity, medium clarity, and roughness. In comparison to traditional PEDOT-PPS, PPS with a conductive polymer, such as its PANI copolymer solution, offers easier doping, solubility, and film quality for thematic monitors/screens [36]. To create a perforation injection film in two-layer electroluminescence with an orange electroluminescence display in comparison to a single-layer electroluminescence device, PANI self-doping based on aniline and (aminobenzene sulfonic acid) is assembled in its ITO glass [37]. Similar to ZnO/PANI nanowire (type n/p), this organic-mineral ray-emitting diode is produced by the recombination of the electron boundary in the conduction band and the holes over a broad light spectrum [38]. The less the separation between the heterostructure of hole transport layer and emissive layer the greater will be the efficiency of the device due to the more transport of holes from HTL to EL in the device. Similarly, the less the separation between the heterostructure of electron transport layer and emissive layer the greater will be the efficiency of the device due to the more transport of electrons from ETL to EL in the device (**Figure 10**).

4.2.4 Waste-water treatment applications

Application of PANI based binary nanocomposites as absorbents in waste water treatment: Polyaniline nanocomposites have been employed as adsorbents to remove various impurities from effluent in recent years. Due to their excellent interaction

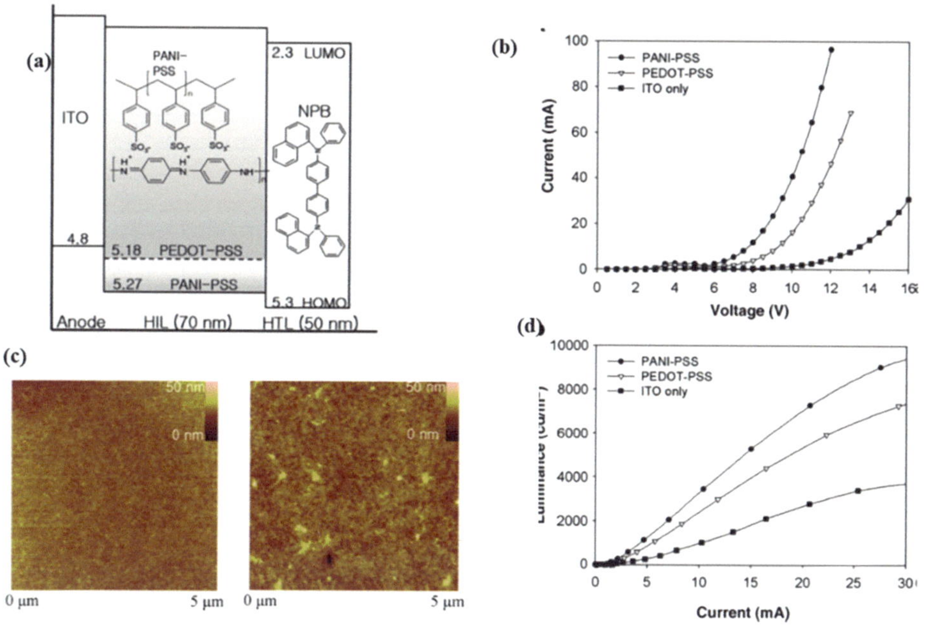

Figure 10.
(a) energy level diagram of heteromolecular (b) Current versus voltage curve of OLED device (c) AFM surface structure of layer and (d) luminescence versus current curve of device, Image reused with permission from Ref. [37] Copyright 2007 Elsevier.

Adsorbent	Pollutant	Langmuir maximum capacity (q_{max}) in mg/g	pH	Conc. (mg/L)	Adsorbent dose (mg)	Ref.
PANI/PVP	Mn (II)	50.30	7	100	250	[39]
PANI/Fe	CR	99.6	7	100	1000	[40]
PANI/zeolite	Cr (VI)	—	2	50	200	[41]
PANI/ $ZnFe_2O_4$	Rhodamine BRHB	1000	2	10	500	[42]
PANI/PAN	Cr (VI)	67.3	2	5	10	[43]

Table 2.
Contains the PANI based nanocomposite employed as an absorbent of pollutants.

with the functional groups of PANI nanocomposites, further research has been conducted on the adsorption of organic pigments and ions of heavy metals. In the study of the association between PANI nanocomposites and impurities, various adsorption parameters including pH, incubation time, adsorbent dose, temperature, adsorbent nature, and pollutant concentration are investigated. The different PANI based nanocomposite employed as an absorbent of pollutants are listed in **Table 2**. To confirm the potential of PANI nanocomposites as adsorbents for water purification, the efficacy and adsorption capability of nanocomposites can be estimated based on these parameters.

5. Conclusions

Polyaniline (PANI) is a conducting polymer with a wide range of applications due to its unique electrical, mechanical, and chemical properties. PANI can be synthesized by various methods, including chemical oxidation, electrochemical polymerization, and template-assisted synthesis. Polyaniline exists in different oxidation states: emeraldine base (EB), emeraldine salt (ES), and pernigraniline (PN). The EB form is most widely studied and has both conductive and insulating properties. The doping of polyaniline with acids or other dopants enhances its electrical conductivity and other properties. Polyaniline exhibits good thermal stability, high mechanical strength, and excellent environmental stability. Polyaniline has wide range of applications such as in conductive coating material for corrosion protection of metals, anti-static coatings, and electromagnetic shielding. Polyaniline has shown promise in applications such as supercapacitors and rechargeable batteries due to its high specific capacitance and good cycling stability. Polyaniline-based sensors are used for gas sensing (e.g., detection of ammonia, carbon monoxide), humidity sensing, and biosensing applications. Polyaniline-based actuators can convert electrical energy into mechanical motion, making them suitable for applications in robotics and microfluidics. Polyaniline can be employed in organic field-effect transistors (OFETs), organic light-emitting diodes (OLEDs), and other organic electronic devices. Also, polyaniline-based adsorbents have been explored for the removal of various pollutants from water, including heavy metals and organic contaminants. Advancements in the design of polyaniline-based nanocomposites and hybrid materials are expected to further enhance its performance in various fields. The development of scalable and cost-effective synthesis methods is crucial to facilitate the commercialization of polyaniline-based technologies. It's worth noting that the field of polyaniline research is vast, and ongoing discoveries may expand the range of applications and understanding of its properties.

Acknowledgements

The author sincerely than IIT (ISM) Dhanbad for all assistance and support.

Conflict of interest

The authors declare no conflict of interest.

Author details

Gobind Mandal*, Jayanta Bauri, Debashish Nayak, Sanjeev Kumar, Sarfaraz Ansari and Ram Bilash Choudhary*
Nanostructured Composite Materials Laboratory (NCML), Department of Physics, Indian Institute of Technology (Indian School of Mines), Dhanbad, Jharkhand, India

*Address all correspondence to: gobindism@gmail.com and rbcism@gmail.com

References

[1] Bhadra S, Khastgir D, Singha NK, Lee JH. Progress in preparation, processing and applications of polyaniline. Progress in Polymer Science. 2009;**34**:783-810. DOI: 10.1016/j.progpolymsci.2009.04.003

[2] Lu H, Li X, Lei Q. Conjugated conductive polymer materials and its applications: A mini-review. Frontiers in Chemistry. 2021;**9**. DOI: 10.3389/fchem.2021.732132

[3] Bouarissa A, Gueddim A, Bouarissa N, Djellali S. Band structure and optical properties of polyaniline polymer material. Polymer Bulletin. 2018;**75**:3023-3033. DOI: 10.1007/s00289-017-2189-6

[4] Shirakawa H, McDiarmid A, Heeger A. Focus article: Twenty-five years of conducting polymers. Chemical Communications. 2003;**1**:1-4. DOI: 10.1039/b210718j

[5] Bauri J, Choudhary RB, Mandal G. Recent advances in efficient emissive materials-based OLED applications: A review. Journal of Materials Science. 2021;**56**:18837-18866. DOI: 10.1007/s10853-021-06503-y

[6] Kumar R, Ansari MO, Parveen N, Oves M, Barakat MA, Alshahri A, et al. Facile route to a conducting ternary polyaniline@TiO 2/GN nanocomposite for environmentally benign applications: Photocatalytic degradation of pollutants and biological activity. RSC Advances. 2016;**6**:111308-111317. DOI: 10.1039/c6ra24079h

[7] Babel V, Hiran BL. A review on polyaniline composites: Synthesis, characterization, and applications. Polymer Composites. 2021;**42**:3142-3157. DOI: 10.1002/pc.26048

[8] Zarrintaj P, Ahmadi Z, Vahabi H, Ducos F, Reza Saeb M, Mozafari M. Polyaniline in retrospect and prospect. Materials Today: Proceedings. 2018;**5**:15852-15860. DOI: 10.1016/j.matpr.2018.05.084

[9] Mandal G, Choudhary RB. MnO2 integrated emeraldine polyaniline (PANI-MnO2) nanocomposites with inflated opto-electricaltraitsas ETLs for OLED applications. Materials Science in Semiconductor Processing. 2022;**151**:107000. DOI: 10.1016/j.mssp.2022.107000

[10] Sai S, Kumar A. Synthesis and morphological study of polyaniline. European Journal of Molecular and Clinical Medicine. 2020;**07**:2020

[11] Wang B, Tang J, Wang F. Electrochemical polymerization of aniline. Synthetic Metals. 1987;**18**:323-328. DOI: 10.1016/0379-6779(87)90899-X

[12] Lawal AT, Wallace GG. Vapour phase polymerisation of conducting and non-conducting polymers: A review. Talanta. 2014;**119**:133-143. DOI: 10.1016/j.talanta.2013.10.023

[13] Winther-Jensen B, West K. Vapor-phase polymerization of 3,4-Ethylenedioxythiophene: A route to highly conducting polymer surface layers. Macromolecules. 2004;**37**:4538-4543. DOI: 10.1021/ma049864l

[14] Majeed AH, Mohammed LA, Hammoodi OG, Sehgal S, Alheety MA, Saxena KK, et al. A review on polyaniline: Synthesis, properties, nanocomposites, and electrochemical applications. International Journal of Polymer Science. 2022;**2022**. DOI: 10.1155/2022/9047554

[15] Lee JH, Prud'homme RK, Aksay IA. Cure depth in photopolymerization: Experiments and theory. Journal of Materials Research. 2001;**16**:3536-3544. DOI: 10.1557/JMR.2001.0485

[16] Maponya TC, Hato MJ, Somo TR, Ramohlola KE, Makhafola MD, Monama GR, et al. Polyaniline-based nanocomposites for environmental remediation. In: Trace Met. Environ.—New Approaches Recent Adv. London, U.K.: IntechOpen; 2021. DOI: 10.5772/intechopen.82384

[17] Zarrintaj P, Vahabi H, Saeb MR, Mozafari M. Application of polyaniline and its derivatives. In: Fundam. Emerg. Appl. Polyaniline. Amsterdam, Netherlands: Elsevier; 2019. pp. 259-272. DOI: 10.1016/B978-0-12-817915-4.00014-2

[18] Ansari S, Bilash R, Gupta A. Nanoflower copper sulphide intercalated reduced graphene oxide integrated polypyrrole nano matrix as robust symmetric supercapacitor electrode material. Journal of Energy Storage. 2023;**59**:106446. DOI: 10.1016/j.est.2022.106446

[19] Vangari M, Pryor T, Jiang L. Supercapacitors: Review of materials and fabrication methods. Journal of Energy Engineering. 2013;**139**:72-79. DOI: 10.1061/(asce)ey.1943-7897.0000102

[20] Snook GA, Kao P, Best AS. Conducting-polymer-based supercapacitor devices and electrodes. Journal of Power Sources. 2011;**196**:1-12. DOI: 10.1016/j.jpowsour.2010.06.084

[21] Choudhary RB, Ansari S, Purty B. Robust electrochemical performance of polypyrrole (PPy) and polyindole (PIn) based hybrid electrode materials for supercapacitor application: A review. Journal of Energy Storage. 2020;**29**:101302. DOI: 10.1016/j.est.2020.101302

[22] Zhang X, Ji L, Zhang S, Yang W. Synthesis of a novel polyaniline-intercalated layered manganese oxide nanocomposite as electrode material for electrochemical capacitor. Journal of Power Sources. 2007;**173**:1017-1023. DOI: 10.1016/j.jpowsour.2007.08.083

[23] Khdary NH, Abdesalam ME, Enany GEL. Mesoporous polyaniline films for high performance supercapacitors. Journal of The Electrochemical Society. 2014;**161**:63-68. DOI: 10.1149/2.0441409jes

[24] Wang H, Lin J, Shen ZX. Polyaniline (PANi) based electrode materials for energy storage and conversion. Journal of Science: Advanced Materials and Devices. 2016;**1**:225-255. DOI: 10.1016/j.jsamd.2016.08.001

[25] Shafi PM, Ganesh V, Bose AC. LaMnO 3/RGO/PANI ternary nanocomposites for supercapacitor electrode application and their outstanding performance in all-solid-state asymmetrical device design. ACS Applied Energy Materials. 2018;**1**:2802-2812. DOI: 10.1021/acsaem.8b00459

[26] Ma Y, Zhao D, Chen Y, Huang J, Zhang Z, Zhang X, et al. A novel core-shell polyaniline/graphene oxide/copper nanocomposite for high performance and low-cost supercapacitors. Chemical Papers. 2019;**73**:119-129. DOI: 10.1007/s11696-018-0556-x

[27] Mu B, Zhang W, Shao S, Wang A. Glycol assisted synthesis of graphene–MnO2–polyaniline ternary composites for high performance supercapacitor electrodes. Physical Chemistry Chemical Physics. 2014;**16**:7872. DOI: 10.1039/c4cp00280f

[28] Purty B, Choudhary RB, Kandulna R, Singh R. Remarkable enhancement in electrochemical capacitance value of Ag-ZnO/PANI composite for supercapacitor application. AIP Conference Proceedings. 2019;**2115**:030588. DOI: 10.1063/1.5113427

[29] Xiong P, Hu C, Fan Y, Zhang W, Zhu J, Wang X. Ternary manganese ferrite/graphene/polyaniline nanostructure with enhanced electrochemical capacitance performance. Journal of Power Sources. 2014;**266**:384-392. DOI: 10.1016/j.jpowsour.2014.05.048

[30] Hosseinnezhad M, Gharanjig K, Moradian S, Saeb MR. In quest of power conversion efficiency in nature-inspired dye-sensitized solar cells: Individual, co-sensitized or tandem configuration? Energy. 2017;**134**:864-870. DOI: 10.1016/j.energy.2017.06.045

[31] Hosseinnezhad M, Saeb MR, Garshasbi S, Mohammadi Y. Realization of manufacturing dye-sensitized solar cells with possible maximum power conversion efficiency and durability. Solar Energy. 2017;**149**:314-322. DOI: 10.1016/j.solener.2016.11.011

[32] Liu B, Huo L, Si R, Liu J, Zhang J. A general method for constructing two-dimensional layered mesoporous mono- and binary-transition-metal nitride/graphene as an ultra-efficient support to enhance its catalytic activity and durability for electrocatalytic application. ACS Applied Materials & Interfaces. 2016;**8**:18770-18787. DOI: 10.1021/acsami.6b03747

[33] Zhang D, Ryu K, Liu X, Polikarpov E, Ly J, Tompson ME, et al. Transparent, conductive, and flexible carbon nanotube films and their application in organic light-emitting diodes. Nano Letters. 2006;**6**:1880-1886. DOI: 10.1021/nl0608543

[34] Wang G, Xing W, Zhuo S. The production of polyaniline/graphene hybrids for use as a counter electrode in dye-sensitized solar cells. Electrochimica Acta. 2012;**66**:151-157. DOI: 10.1016/j.electacta.2012.01.088

[35] Jang J, Ha J, Kim K. Organic light-emitting diode with polyaniline-poly(styrene sulfonate) as a hole injection layer. Thin Solid Films. 2008;**516**:3152-3156. DOI: 10.1016/j.tsf.2007.08.088

[36] Huh DH, Chae M, Bae WJ, Jo WH, Lee T-W. A soluble self-doped conducting polyaniline graft copolymer as a hole injection layer in polymer light-emitting diodes. Polymer (Guildf). 2007;**48**:7236-7240. DOI: 10.1016/j.polymer.2007.09.046

[37] Yang C-H, Chih Y-K. Molecular assembled self-doped polyaniline interlayer for application in polymer light-emitting diode. The Journal of Physical Chemistry. B. 2006;**110**:19412-19417. DOI: 10.1021/jp0612174

[38] Liu YY, Wang XY, Cao Y, Chen XD, Xie SF, Zheng XJ, et al. A flexible blue light-emitting diode based on ZnO nanowire/polyaniline heterojunctions. Journal of Nanomaterials. 2013;**2013**:1-4. DOI: 10.1155/2013/870254

[39] Hallajiqomi M, Eisazadeh H. Adsorption of manganese ion using polyaniline and it's nanocomposite: Kinetics and isotherm studies. Journal of Industrial and Engineering Chemistry. 2017;**55**:191-197. DOI: 10.1016/j.jiec.2017.06.045

[40] Bhaumik M, McCrindle RI, Maity A. Enhanced adsorptive degradation of Congo red in aqueous solutions using polyaniline/Fe0 composite nanofibers. Chemical Engineering Journal. 2015;**260**:716-729. DOI: 10.1016/j.cej.2014.09.014

[41] Shyaa AA, Hasan OA, Abbas AM. Synthesis and characterization of polyaniline/zeolite nanocomposite for the removal of chromium(VI) from aqueous solution. Journal of Saudi Chemical Society. 2015;**19**:101-107. DOI: 10.1016/j.jscs.2012.01.001

[42] Rachna K, Agarwal A, Singh N. Preparation and characterization of zinc ferrite—Polyaniline nanocomposite for removal of rhodamine B dye from aqueous solution. Environmental Nanotechnology, Monitoring & Management. 2018;**9**:154-163. DOI: 10.1016/j.enmm.2018.03.001

[43] Ren J, Huang X, Wang N, Lu K, Zhang X, Li W, et al. Preparation of polyaniline-coated polyacrylonitrile fiber mats and their application to Cr(VI) removal. Synthetic Metals. 2016;**222**:255-266. DOI: 10.1016/j.synthmet.2016.10.027